ADVANCES AND TRENDS IN GEODESY, CARTOGRAPHY AND
GEOINFORMATICS II

T0203862

Advances and Trends in Geodesy, Cartography and Geoinformatics

ISSN: 2643-2420
e-ISSN: 2643-2404

ABOUT THE SERIES
The Book Series "Advances and Trends in Geodesy, Cartography and Geoinformatics" is, in line with its long tradition, devoted to the publication of proceedings of peer-reviewed international conferences focusing on presenting technological and scientific advances in modern geodesy, geoinformatics, cartography, photogrammetry, remote sensing, geography, and related sciences. The implementation of the research findings presented in the Book Series plays an extremely important role in accelerating the development of all these disciplines. The aim of the Series is to stimulate advanced education and training through the wide dissemination of new scientific knowledge and trends in Geodesy, Cartography and Geoinformatics to a broad group of scientists and specialists.

PROCEEDINGS OF THE 11TH INTERNATIONAL SCIENTIFIC AND TECHNICAL CONFERENCE ON GEODESY, CARTOGRAPHY AND GEOINFORMATICS (GCG 2019), 10—13 SEPTEMBER, 2019, DEMÄNOVSKÁ DOLINA, LOW TATRAS, SLOVAKIA

Advances and Trends in Geodesy, Cartography and Geoinformatics II

Editors

Soňa Molčíková, Viera Hurčíková & Peter Blišťan

Institute of Geodesy, Cartography and Geographical Information Systems Technical University of Košice, Slovakia

CRC Press
Taylor & Francis Group
Boca Raton   London   New York

CRC Press is an imprint of the
Taylor & Francis Group, an **informa** business

A BALKEMA BOOK

CRC Press/Balkema is an imprint of the Taylor & Francis Group, an informa business

© 2020 Taylor & Francis Group, London, UK

Typeset by Integra Software Services Pvt. Ltd., Pondicherry, India

Published by: CRC Press/Balkema
      Schipholweg 107C, 2316XC Leiden, The Netherlands
      e-mail: Pub.NL@taylorandfrancis.com
      www.crcpress.com – www.taylorandfrancis.com

ISBN: 978-0-367-34651-5 (Hbk)
ISBN: 978-0-367-51581-2 (pbk)
ISBN: 978-0-429-32702-5 (eBook)
DOI: https://doi.org/10.1201/9780429327025

Advances and Trends in Geodesy, Cartography and Geoinformatics II –
Molčíková, Hurčíková, & Blišťan (eds)
© 2020 Taylor & Francis Group, London, ISBN 978-0-367-34651-5

Table of contents

Advances and Trends in Geodesy, Cartography and Geoinformatics II –
Molčíková, Hurčíková, & Blišťan (eds)
© 2020 Taylor & Francis Group, London, ISBN 978-0-367-34651-5

Table of contents

Preface

The anniversary 11th International Scientific and Technical Conference on Geodesy, Cartography and Geoinformatics (GCG 2019) was organized under the auspices of the Faculty of Mining, Ecology, Process Control and Geotechnologies, Technical University of Košice, Slovakia. The co-organizers of this conference were the Faculty of Science of the Pavol Jozef Šafárik University in Košice (Slovakia), the Faculty of Civil Engineering of Slovak University of Technology in Bratislava (Slovakia), the Faculty of Civil Engineering of Czech Technical University in Prague (Czech Republic), the Kielce University of Technology (Poland), AGH University of Science and Technology in Krakow (Poland), Miskolci Egyetem (Hungary), Upper Nitra Mines Prievidza, plc. (Slovakia) and the Slovakian Mining Society (Slovakia). The conference was held September 10 – 13, 2019 in the Low Tatras, Slovakia.

The purpose of the conference was to facilitate a meeting that would present novel and fundamental advances in the field of geodesy, cartography, geoinformatics, real estate, cadastre and land consolidation for scientists, researchers, and professionals around Middle Europe. Conference participants had the opportunity to exchange and share their experiences, research, and results with other colleagues, experts, and professionals. Participants also had the possibility to extend their international contacts and relationships for furthering their activities.

The conference focused on a wide spectrum of topics and areas of geodesy, cartography, geoinformatics and real estate as listed below:

1. Surveying and mine surveying
   - Geodetic networks, data processing
   - Engineering surveying and deformation measurement
   - Photogrammetry and remote sensing
   - Real estate cadastre and land consolidation

2. Geodetic control and geodynamics
   - Cosmic and satellite geodesy, theory and applications
   - Vertical reference systems
   - Absolute and relative gravimetry

3. Cartography, geoinformatics and real estate
   - Collection and processing of spatial data,
   - Data standards, infrastructure, metadata, geodatabases
   - Spatial analyses and modeling
   - Development and application of methods and models for spatial processes
   - Digital cartographic systems
   - 3D visualization of spatial data, publishing data on the internet, virtual reality
   - Real estate, cadastre and land consolidation

30 papers from countries V4 were accepted for publication in the conference proceedings of the Conference. Each of the accepted papers was reviewed by selected reviewers in accordance with the scientific area and orientation of the papers.

The editors would like to express their many thanks to the members of the Organizing and Scientific Committees. The editors would also like to express a special thanks to all reviewers, sponsors and conference participants for their intensive cooperation to make this conference successful.

Soňa Molčíková

Viera Hurčíková

Peter Blišťan

Advances and Trends in Geodesy, Cartography and Geoinformatics II –
Molčíková, Hurčíková, & Blišťan (eds)
© 2020 Taylor & Francis Group, London, ISBN 978-0-367-34651-5

Committees of GCG 2019

Scientific Committee

Prof. Jaroslav HOFIERKA
Pavol Jozef Šafárik University in Košice, Slovakia

Prof. Juraj JANÁK
Slovak University of Technology in Bratislava, Slovakia

Prof. Alojz KOPÁČIK
Slovak University of Technology in Bratislava, Slovakia

Prof. Janka SABOVÁ
Technical University of Košice, Slovakia

Prof. Jacek SZEWCZYK
Polish Polytechnic of Świętokrzyskie in Kielce, Poland

Prof. Martin ŠTRONER
Czech Technical University in Prague, Czech Republic

Prof. Ján TUČEK
Technical University in Zvolen, Slovakia

Assoc. Prof. Peter BLIŠŤAN
Technical University of Košice, Slovakia

Assoc. Prof. Renata ĎURAČIOVÁ
Slovak University of Technology in Bratislava, Slovakia

Assoc. Prof. Marek FRAŠTIA
Slovak University of Technology in Bratislava, Slovakia

Assoc. Prof. Jana IŽVOLTOVÁ
University of Žilina, Slovakia

Assoc. Prof. Radovan MACHOTKA
Brno University of technology, Czech Republic

Assoc. Prof. Katarína PUKANSKÁ
Technical University of Košice, Slovakia

Assoc. Prof. Rudolf URBAN
Czech Technical University in Prague, Czech Republic

Dr. Endre DOBOS
Miskolci Egyetem, Hungary

Dr. Branislav DROŠČÁK
Geodetic and Cartographic Institute Bratislava, Slovakia

Dr. Piotr PARZYCH
AGH University of Science and Technology in Krakow, Poland

MSc. Ivan HORVÁTH
Geodetic and Cartographic Institute Bratislava, Slovakia

Organizing Committee

Dr. Soňa Molčíková
Dr. Viera Hurčíková
Dr. Vladislava Zelizňaková

List of Reviewers

K. Bartoš, M. Bajtala, F. Beneš, A. Bieda, M. Bindzárová Gergeľová, P. Blišťan, D. Bobíková, O. Čerba, R. Ďuračiová, R. Fencík, P. Frąckiewicz, M. Fraštia, J. Gajdošík, M. Gallay, J. Gašinec, J. Hofierka, V. Hurčíková, J. Ižvoltová, J. Janák, P. Jašek, J. Kaňuk, A. Kopáčik, Ľ. Kovanič, Ž. Kuzevičová, S. Labant, R. Machotka, E. Mičietová, M. Mojzeš, S. Molčíková, J. Papčo, J. Pokorný, K. Pukanská, Š. Rákay, J. Sabová, J. Szewczyk, M. Štroner, A. Švejda, M. Talich, R. Urban, B. Veverka, G. Weiss, P. Zahorec

Surveying and mine surveying

Advances and Trends in Geodesy, Cartography and Geoinformatics II –
Molčíková, Hurčíková, & Blišťan (eds)
© 2020 Taylor & Francis Group, London, ISBN 978-0-367-34651-5

Comparison of different measurement methods of crane runway

J. Braun, H. Fladrova & K. Prager
Department of Special Geodesy, Faculty of Civil Engineering, CTU in Prague, Prague, Czech Republic

ABSTRACT: The aim of the paper is to introduce the surveying and evaluating the historic crane runway in the Karany water treatment plant. The crane runway was built in 1911 and is operated to the present without any major modifications. The aim of the geodetic survey was the control of the directional and vertical straightness of rails and the rail gauge. Accuracy analyzes and deviations evaluation were performed in accordance with technical standard CSN 73 5130. The runway was measured by two different methods. The first method was measurement in the local geodetic network with three standpoints. Automatic targeting to prisms was used, and the least squares adjustment was performed. The second method of measurement was laser scanning by multistation Leica MS60 from 14 standpoints and then evaluation of the point cloud and determination the parameters of the crane runway. Comparable results were obtained from both methods.

1 INTRODUCTION

The crane runway is a transport facility in an industrial plant and is used to move bulky and heavy loads. The crane moves on rails, which can be on the ground, pillars or suspended on consoles. The correct geometric position of rails is an essential element in safety, functionality and economy of operation. Geodetic methods monitor the straightness of individual rails, rail gauge and rail elevation. Regular checks should be carried out to avoid damage and accidents (Filipiak-Kowszyk & Kaminski 2016, Vašková et al. 2017). Errors in rail gauge cause excessive lateral force on the crane and height errors between the rails would result to uneven distribution of weight (Shortis & Ganci 1995). There are technical standards for determining geometric parameters that prescribe the maximum possible deviation from the ideal state. In the Czech system, these are the ČSN 73 5130 (1994) and ČSN ISO 12488-1 (2012) standards, which are linked to the international system of technical standards.

The rail is the main object of measurement according to technical standards. At the prescribed points, the position relative to the geometric center of the rail and the height to the top of the rail are determined. Points on the rail suitable for measurement varies according to the type and construction of the crane runway itself. Furthermore, the position of the bumper is determined, and in special cases the position of the crane itself. Measurements should be taken when the crane is stopped in one extreme position so that the measurement is not affected by its load.

Standard measurement procedures have long been established and sufficient. With the development of technology, there are new procedures that can be applied mainly in special cases when the crane runway is difficult to access. Methods for measuring geometry parameters of rail can also be selected according to size of crane runway and accuracy requirements.

The simplest and oldest method is the line of sight in combination with leveling. This method is particularly suitable for short and easily accessible crane runway (Pospíšil 2016). Another historical method of measurement is the semi-polar method. In this method, angular measurements are used on all points and length measurements are used only on some points. The lengths needed to determine the lateral deviations are calculated using trigonometric solutions. The method has been widely used before the advent of accurate distance meters in the total station.

Currently, the most widely is used spatial polar method. The precise total stations allow to determine all the elements at once. Most of today's total stations are equipped with a distance meter with a precision of a few millimeters. To maintain the accuracy given by the technical standards, it is necessary to stabilize the standpoint so that a length error is projected into the rail gauges as little as possible. If the standpoint is stabilized between the rails so as to form an approximately isosceles triangle with the measured gauges, then the direction determination error is projected to a greater extent than the length error. The direction error is of the order of magnitude less than the length error (Štroner et al. 2010).

For larger crane runways and high-accuracy requirements runways, the method of spatial geodetic network is used, which is usually measured from multiple standpoints simultaneously to individual points. Additional methods are photogrammetry and laser scanning (Blistan & Kovanic 2014). From point clouds can be analyzed the overall course of rails. It is also possible to focus on parts that are not measured by classical methods (Křemen et al. 2008, Kregar et al. 2017, Kovanic et al. 2019). The development of technology is aimed at automated measurement and dynamic tracking of crane runway using special moving constructions that perform continuous measurement (Kyrinovič & Kopáčik 2010, Dennig et al. 2017).

This paper focuses on the crane runway in the Karany water treatment plant (near Prague), which was built in 1911 and is still operated without major repairs. The aim of this work was to check the geometrical parameters of the runway and test the possibilities of the multistation Leica MS60. The data were measured as part of the diploma thesis (Prager 2019). The crane runway is difficult to access and therefore the less common geodetic network method was chosen and laser scanning was the alternative method.

2   CRANE RUNWAY

The crane was manufactured in 1911 and is used to manipulate the pump during the fault and further to move heavy and large objects. It is a unique riveted construction that still suits current operation. Output to crane and crane runway is secured by ladder with protective cage. There are no revision bridges around the crane runway to allow safe access to the rail.

Figure 1.   Crane runway in Karany water treatment.

The crane runway has a length of 74.3 m and is fitted with a 60 x 40 mm rail, which lies on an iron beam of the "I" profile. Rails are located 8 m above the engine room floor and the designed rail gauge is 23.810 m. Each rail is supported by 15 columns. The columns are 5 m apart.

3   GEODETIC NETWORK

3.1   *Accuracy analysis*

Determining the geometric parameters of crane runway by measuring in geodetic network is not a new procedure (Shortis & Ganci 1995, Štroner et al. 2010). However, accuracy analyzes are often neglected. Measurement of the network can be done in various ways, and not always satisfactory results are obtained which meet the high accuracy requirements.

Accuracy parameters given by the technical standard were the initial requirement for precision of measuring and configuration of standpoints. The tolerance values of rail gauge 7.7 mm and rail elevation 12.0 mm were used. The required standard deviations of rail gauge 0.8 mm and elevation between two points 1.2 mm were determined. Values are based on the formula:

$$\sigma_{required} = \frac{0.2 \cdot T}{u_p} \tag{1}$$

where $\sigma_{required}$ = required standard deviation; T = tolerance; and u_p = 2 - normalized value of normal distribution.

Software PrecisPlanner 3D was used for planning configuration of standpoints and measurement repeats (Štroner 2010). The software allows to determine the accuracy of the resulting coordinates based on approximate coordinates, selection of measured values and their precision. It also determines the covariance matrix that allows further calculations of the precision of derived values (eg lengths, angles).

The accuracy of the total station being considered for direction 0.3 mgon and 1 mm for the length, measured in the two faces of the telescope. Because there are no safety bridges on the crane runway, there has been an effort to minimize the movement of the surveyor at height. There are 17 profiles (34 points) on the runway. There are 15 profiles above the supporting columns. There is one profile at each end of the bumper. Three measurement configurations have been designed that meet the precision requirements.

The first configuration was designed to be measured from two standpoints on the floor. The standpoints were near the walls in the middle of the runway and from each was measured on the opposite rail and connecting points. In the second configuration, measurements were again done from the floor. Four standpoints were proposed, which were around the corners of the hall. Again, the opposite rail and connecting points were measured. These variants required repetitive movement of the surveyor on the rail and were therefore rejected. The third configuration has been designed so that there is standpoint at the end of each rail and is measured by two instruments at the prescribed points and at the connecting points. The third standpoint on the floor was designed to measure the end of the runway, from which only two profiles and connecting points were measured. The resulting average projected standard deviations are 0.5 mm in position and 0.2 mm in height. By law of propagation of standard deviations, the standard deviation of gauge and height difference were derived. The projected values were less than the required value and the expected precision was met.

3.2   *Measurement of network*

The first step was to build a reference network - connecting points. Points were selected on the floor using sharp corner of tiles (10 points) and also on the walls using cross mark (4 points). The signalization was made with the Leica GMP111 mini-prism, which was placed in the rack. The reference network was designed so that at least 4 points could be seen from any point in the hall and the resection could be performed. The reference points were determined from one standpoint in one measurement group using the Leica MS60 with automatic targeting (standard deviation of direction 0.3 mgon, standard deviation of distance 1 mm).

The instruments Trimble S6 HP and Trimble S8 HP (standard deviation of direction 0.3 mgon, standard deviation of distance 1 mm) were used for measurements on rails. The instruments can be operated remotely using a controller. The crane was shut down at end of runway. Special handles for the total station were attached to the end of the rails. Total stations were carefully attached and carefully levelling of the instrument was performed. Each instrument done resection to 5 points of the reference network and standard deviation of coordinate less than 1 mm was achieved. The surveyor on the rail gradually shifted along the prescribed points. The points were set in advance with a step of 5 m so that they were always above the center of the support column. The position of the point was set out so that the points on rails were facing each other and the rail gauge could be determined directly from coordinates. The center of the rail was always marked with a centering shear. Point signalization was performed with a Leica GMP101 mini-prism. Both total stations were measured at once on the target in both telescope faces. There were measured 15 profiles. Then the instruments were brought to the floor. The crane was moved to the center of the hall. On the floor, a third standpoint with the Trimble S8 HP was on the tripod. The resection on 5 points was done. 2 profiles in the crane parking space were measured. The Leica MS60 was not used because a controller was not available to control the instrument remotely.

Figure 2.   Measurement on rail with Trimble S6 HP.

3.3   *Evaluation of measurement*

The output of the measurement were the field books, which were processed in the EASYNet software, which serves for adjustment of geodetic networks. The software carries out automatic field book processing, estimation about accuracy of measurement, and robust adjustment at which it automatically excludes outlier measurements. The network was adjusted as fixed, the points of the reference network were fixed. In this case there were 237 measured values. 20 outliers were eliminated in adjustment. The average standard deviation of the coordinates was 0.3 mm (max. 0.8 mm). Accuracy assessment is performed using a priori and a posteriori standard deviation, which indicates the consistency or mismatch of the standard deviations that are considered and actually achieved in the measurement. The a priori standard deviation was 1. The a posterior standard deviation was 0.72. This means that the real accuracy of the measurements was better than the expected accuracy. The results of the adjustment are the coordinates of 36 points and their standard deviations. The rail gauge and the elevations were calculated from the coordinates.

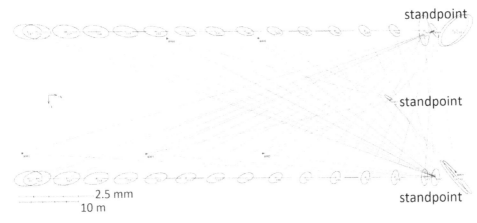

standpoint

standpoint

standpoint

2.5 mm
10 m

Figure 3. Position of points, ellipse of errors and measurement vectors.

4 LASER SCANNING

4.1 Measurement with Leica MS60

The laser scanning method is usually used so that the standpoints are at the end of the runway and the individual runway segments are scanned sequentially (Kregar et al. 2017). Another way is that the scanner moves over the revision bridges and scans individual runway segments. The scanner positions are then linked using identical points and transformed into a coordinate system in a suitable software. For this measurement, it was proposed to use the features and capabilities of the multi-station Leica MS60. The instrument allows to calculate the stand-point coordinates, then the measured point cloud is oriented in the reference coordinate system. The resulting point cloud can be immediately compared to the values obtained by the measurement in the geodetic network.

The total station was placed on the bridge of the crane runway. The bridge was gradually moved to 7 positions (10 m interval). At each bridge position, the instrument was gradually set up over both rails - 14 standpoints. The instrument always done resection at least on 4 points of the reference network. The standard deviations of the standpoints coordinates were less than 2 mm. A section of the rail about 10 m length was scanned from each standpoint. The section was chosen so that it was not affected by the weight of the bridge. Profiles were scanned at 5 m - 15 m from the bridge. The resolution of point cloud was 4 mm to 10 m. Only the narrow rail strip defined on the instrument's display was scanned. The measurement took about 15 minutes for each standpoint. A supplementary point cloud was measured from the floor for a better orientation in the hall.

4.2 Evaluation of measurement

Point clouds were cleaned and then the sections were spaced with planes (vertical on the out-side and horizontal on the upper surface) in software Cyclone. The intersection of the planes then formed a line that defined the course of the rail. The sections were chosen among the pillars. The deviation from the spaced plane did not exceed 2 mm. If the individual lines did not follow each other, a connecting center point to both lines was found. The result was a drawing of two lines that represented the inner edges of the rails. To determine the gauge, it was necessary to correct the spacing by a rail thickness of 60 mm.

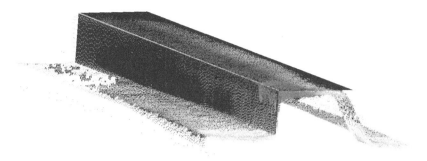

Figure 4. Cleared point cloud spaced by planes.

5 CONCLUSION

Individual gauges and elevations were calculated from both measurement methods. The maximum gauge difference is 7.7 mm. This value was exceeded in two cases (maximum deviation of 10.2 mm). The elevation change limit is 12.0 mm. This value was exceeded in 4 cases. The maximum elevation is 28.5 mm. According to these results, the crane runway should be rectified.

Both results of measurement were compared. Similar results were obtained from both methods. The course of deviations is the same. The differences are only in the deviation sizes when some values differed by 2 mm - 5 mm (10% of the data) against the values from the geodetic network.

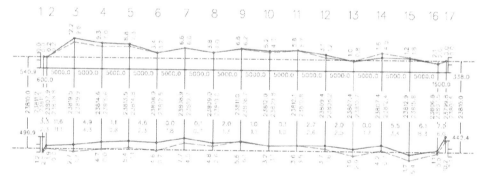

Figure 5. Directional deviations – Geodetic network (line), Laser scanning (dashed line).

The measurement time with the Leica MS60 was longer as it was necessary to measure from 14 standpoints. Processing of point cloud was also longer than the geodetic network calculation. The laser scanning method with multi-station is particularly suitable for difficult to access runways and for runways where accuracy is not high. When measuring in a geodetic network, it is necessary to perform the accuracy analysis before the measurement to verify the suitability of the configuration of the standpoints and the accuracy of the instrument.

This work was supported by the Grant Agency of the Czech Technical University in Prague, grant No. SGS19/047/OHK1/1T/11 "Optimization of acquisition and processing of 3D data for purpose of engineering surveying, geodesy in underground spaces and laser scanning".

REFERENCES

Blistan, P. & Kovanic, L. 2014. Application of Selected Geodetic Methods in Opencast Quary with Aim of Their 3d Model. *Inzynieria Mineralna-Journal of the Polish Mineral Engineering Society* (1): 287–298.

ČSN 73 5130 Jeřábové dráhy (Crane Runways). 1994. Český normalizační institut, Czech Republic.

ČSN ISO 12488-1 Jeřáby - Tolerance pro pojezdová kola a pro jeřábové a příčné dráhy - Část 1: Obecně (Cranes - Tolerances for wheels and travel and traversing tracks - Part 1: General). 2012, Úřad pro technickou normalizaci, metrologii a státní zkušebnictví,Czech Republic.

Dennig, D. & Bureick, J. et al. 2017. Comprehensive and Highly Accurate Measurements of Crane Runways, Profiles and Fastenings. *Sensors* vol.17 (1118). ISSN 1424-8220.

Filipiak-Kowszyk, D. & Kaminski, W. 2016. The Application of Kalman Filtering to Predict Vertical Rail Axis Displacements of the Overhead Crane Being a Component of Seaport Transport Structure. *Polish Maritime Research* 2 vol.23: 64–70.

Kovanic, L. & Blistan, P. et al. 2019. Deformation Investigation of the Shell of Rotary Kiln Using Terrestrial Laser Scanning (TLS) Measurement. *Metalurgija* vol. 58(3-4): 311–314.

Kregar, K. & Možina, J. et al. 2017. Control Measurements of Crane Rails Performed by Terrestrial Laser Scanning. *Sensors* vol.17 (1671). ISSN 1424-8220.

Křemen, K. & Koska, B. et al. 2008. Checking of crane rails by terrestrial laser scanning technology. In: *13th FIG Symposium on Deformation Measurement and Analysis*. Lnec – Lisbon.

Kyrinovič, P. & Kopáčik, A. 2010. Automated measurement system for crane rail geometry determination. In: *27th International Symposium on Automation and Robotics in Construction (ISARC 2010)*.

Pospíšil, J. 2016. Měření jeřábových drah pro jejich montáž, revizi a údržbu. *Technická diagnostika*. 2016(2): 15–20.

Prager, K. 2019. Geodetické zaměření jeřábové dráhy v úpravně vody Káraný. Diploma thesis, Faculty of Civil Engineering CTU in Prague, Czech Republic.

Shortis, M.R. & Ganci, G. 1995. Alignment of Crane Rails using a Survey Network. *Australian Surveyor* 40(4).

Štroner, M. & Urban, R. & Třasák, P. 2010. Measuring Unavailable Crane Rails of Method Free Network with Two Standpoints. In: *Zborník článkov z vedecko-odbornej konferencie: GEODÉZIA, KARTOGRAFIA A GEOGRAFICKÉ INFORMAČNÉ SYSTÉMY 2010*. Košice: Technical University BERG Faculty. ISBN 978-80-553-0468-7.

Štroner, M. 2010. Vývoj softwaru na plánování přesnosti geodetických měření PrecisPlanner 3D. *Stavební obzor* vol.19(3): 92–95.

Vašková, V. & Fojtík, R. & Pustka, D. 2017. Analysis of a Crane Runway Failure. *Procedia Engineering* (190): 255–262.

Advances and Trends in Geodesy, Cartography and Geoinformatics II –
Molčíková, Hurčíková, & Blišťan (eds)
© 2020 Taylor & Francis Group, London, ISBN 978-0-367-34651-5

Examination surfaces reclamation areas on the example earth embankment on the campus of the Kielce University of Technology

P. Frąckiewicz & M. Gil

Faculty of Environmental, Geomatic and Energy Engineering, Kielce University of Technology, Kielce, Poland

ABSTRACT: Remedial actions of reclaimed areas rely on the restoration of utility values to areas damaged as a result of human activity. Carrying out activities aimed at restoring natural and non-natural values requires the development of project documentation. Geodesic documentation is an indispensable element. The use of adequate measurement methods and data visualization algorithms supports the process of transforming degraded areas. On the example of embankment on the campus of the Kielce University of Technology, presents methods of graphical and statistical analysis of areas undergoing reclamation. Optimal interpolation algorithms were selected and basic statistical analyzes were performed. Areas under recultivation require geodetic monitoring of all stages of work. Geodetic monitoring supports the determination of the safety of objects within the area's impact. The presented methods of analysis indicate the direction and manner of preparing the geodetic documentation of reclamation areas. The most important element is the optimization of measurement networks for various observation methods and the selection of modeling algorithms for the protection of reclaimed areas.

1 INTRODUCTION

Reclamation is the process of restoring to the original form, giving new utility values or managing in a rational way the areas degraded as a result of human activity. In fact, it consists in improving the physical and chemical properties of the soil with technical and biological methods. Reclamation is a process that requires a series of treatments to restore the degraded area of the former natural value (*Guidelines of the National...*, *2009*).

The geodetic documentation of areas subject to reclamation is the basis for the development of a devastated area development project. After completing the necessary activities covered by the project, it is the basic document presenting the effects of the work. The optimization of geodetic documentation and analysis methods supports the assessment of impact on the surrounding environment.

2 CHARACTERISTICS OF THE TESTED OBJECT

As a research object, the embankment was selected on the campus of the Kielce University of Technology. The object is located in Kielce, the capital of the Świętokrzyskie Province in southern Poland. The location of the research area is presented in Figure 1.

Morphologically, it is an artificial embankment formed around fire protection tanks to protect students' houses and teaching buildings. The slopes of the facility were protected with grass vegetation and low shrubs to prevent the possibility of material runoff.

Figure 1.   Location of the research area (own study).

3   CHARACTERISTICS OF THE TESTED OBJECT

Observations of the surface can be performed by different methods: classical, GNSS satellite, laser scanning or photogrammetric. The result of the last two are point clouds, which, at a properly selected resolution, give an effect close to the actual model of the surface being tested. Although classical methods of measuring do not have the advantage, their application does not exclude the creation of a high-accuracy model. In such cases, the selection of appropriate interpolation algorithms is of great importance.

The methods of observation of the area can be divided due to the number of observed points: point, area or mixed. The point methods are based on measuring previously-substituted points in characteristic cross-sections or points of anticipated changes, and the number of points is small. Area methods consist of measuring regular or irregular grid of points located on the whole surface, and subjective assessment of the selection of location points affects the quality of the generated model. Mixed methods use elements of both methods. The connection allows you to control the area in previously signaled points, and allows you to select the others in such a way that the resulting model is the most faithful reflection of the actual surface (Frąckiewicz et al. 2018).

Observations of the embankment located next to the student house "Bartek" were made using two methods:

1) Classic (total station) made in October 2017 using the STONEX R1 Plus total station (Figure 2a) measuring angles with two-second accuracy. Observations were made in irregular grid of points from 3 points of the local network. The observation set consists of one set of observations consisting of 111 points.

2) Satellite GNSS (GNSS RTN measurement) performed in June 2018 using the modern Sokkia GRX1 receiver (Figure 2b). One set of observations consisting of 125 points. Observations were made in irregular point grids.

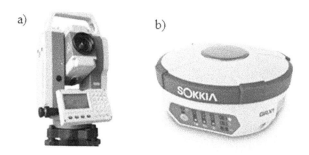

Figure 2.   a) The STONEX R1 plus total station, b) Sokkia GRX1 receiver.

4   GEODETIC METHODS OF ANALYSIS OF RECLAIMED AREAS

Obtained surface observations are only an approximation of its actual course. The use of point methods on the stabilized signs in the case of land recultivated is ineffective, due to the necessity of their removal during the phase of technical reclamation works and the acquisition of information only for these points. Area studies require a large number of observations based on which the coordinates of points are calculated. Coordinates do not have to be determined with high accuracy due to the fact that the whole surface is covered with a dense grid of points - which greatly facilitates the calculation process. This factor is determinative in the case of areas undergoing reclamation due to their considerable range. The development of the obtained data allows for the analysis of models such as the assessment of the object's volume, determination of the inclination angle, terrain profiles and creation of a spatial model of the area. Data analysis supports the selection of the right direction of reclamation and other necessary elements to carry out technical reclamation.

4.1   *Visualization of obtained observations*

Surface modeling can be performed in programs based on different types of interpolation. CAD software uses "classic" methods based on triangulation with linear interpolation. Specialized software (ArcGIS, Petrel, SAGA GIS, Surfer and others) offer more advanced interpolation methods with the possibility of configuring various parameters by the user at the stage of the modeling process (Blistan et al., 2014; Cressie, 1993; Davis, 1986; Hengl, 2009; Namysłowska-Wilczyńska et al., 2006; Surfer books, 2002; Zawadzki, 2015; Zawadzki et al., 2005; Zhou et al., 2007)

The Surfer program was used to analyze the embankment on the campus of the Kielce University of Technology. The program offers 12 interpolation methods based on the GRID model. The methods are based on different algorithms, resulting in a different distribution of predicted values beyond the measurement points. Two most frequently used interpolation methods were used for analysis:

– Kriging (K) - a flexible and accurate method of reticulation based on geostatistics, in which an interpolation error called a kriging variance is determined. The kriging algorithm is effective because it can compensate for the data in the set, giving these areas less weight in the overall forecast. It also allows extrapolation outside the data area. The spatial model of the embankment is presented in Figure 3a.
– Radial Basis Function (RBF) - an interpolation method in which the actual values depend on the distance from the starting point. The method allows to generate a smooth surface from large data sets. Ineffective method for observations with high amplitude of changes of the tested parameter at small distances. The algorithm adopts a specific function and adjusts to the obtained observations. RBF is an alternative to kriging in areas of low variability. The spatial model of the embankment is presented in Figure 3b. The RBF method for the tested area generates an irregular surface incompatible with the actual model.

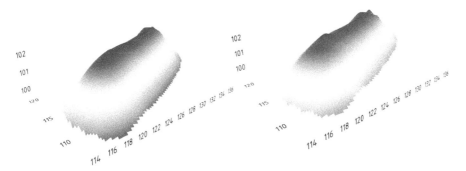

Figure 3. Recultivated embankment models generated by an algorithm a) kriging, b) radial basis function (own study).

4.2 *Analysis of observation results*

Areas undergoing reclamation processes require the development of geodetic documentation in order to present the existing state. In particular, specify the volume of pits shaped at the stage of extraction or embankments created as a result of the storage of gangue and landfills. The determination of the angle of inclination allows, for example, to plan the planting stage, and the terrain profiles present the shape of slope slope relative to the reference level. The applied statistical methods for the example of the embankment are aimed at presenting the data to the required technical documentation.

4.2.1 *Volume determination*

Using the Volume function in the Surfer program allows you to determine the volume of the area relative to a specific reference surface. The user defines the calculation model:

Model 1: observed surface in relation to the horizontal reference plane (Figure 4a)
Model 2: observed surface in relation to another reference surface, eg designed after reclamation (Figure 4b). Different combinations of the location of the reference surface enable the report to obtain adequate positive and negative volumes.

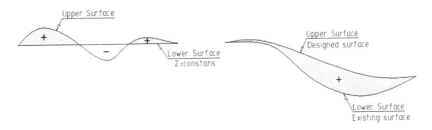

Figure 4. Volume calculation models: a) Model 1: relative to the horizontal reference plane b) Model 2: relative to the other reference reference surface (own study).

The calculation report contains information about projected volume and area measured on the slope. Table 1 summarizes the performed volume comparison for tacheometric measurement and GNSS RTN measurement in relation to the horizontal reference area relative to the lowest point from the observation.

Table 1. The volume of earth masses of the embankment before reclamation (own study).

Model	Upper surface	Lower surface	Positive volume [m³]	Positive surface measured along the slope [m²]
Kriging	Total station		382.7	223.3
	RTN GNSS	Z (const.) =99,5m	381.6	220.7
Radial Basis Function	Total station		382.5	219.8
	RTN GNSS		381.7	218.5

Table 2. Calculated changes in the surface slope angle (own study).

Model	Measurement	Min. inclination [°]	Min. inclination [%]	Max. inclination [°]	Max. inclination [%]	Average inclination [°]	Average inclination [%]	Median [°]	Median [%]
Kriging	Total station	0,3	0,5	60,2	174,6	25,4	47,5	27,2	51,4
	RTN GNSS	0,3	0,5	71,0	290,4	25,9	48,6	27,3	51,6
Radial Basis Function	Total station	0,5	0,9	60,3	175,3	25,0	46,6	27,3	51,6
	RTN GNSS	0,3	0,5	66,6	231,1	25,7	48,1	27,4	51,8

4.2.2 Analysis of the slope inclination

Building embankments and landfills created as a result of mining works after the end of exploitation are characterized by a large inclination angle. It may cause landslides, cause fluvial processes and ground runoff determined mainly by geological structure. Cooperation is needed from the geologist, who will determine the type of substrate and the type of the embankment and in particular determine the angle of internal friction to ensure the stability of the slope. The statistical interpretation of the slope angle in reclamation processes supports the selection of the land development direction.

Determination of area security requires analysis of the existing terrain configuration with respect to the safety configuration, in which the retaining force is greater than the one sliding relative to a determined ground friction angle.

The Surfer program allows you to analyze the slope value of the tested object. The correct interpretation of the minimum, average, maximum and median values allows the assessment of the existing condition. For the tested embankment, the values of slope as a percentage of the slope were determined. In table 2. the results obtained are summarized.

4.2.3 Terrain profiles

Graphical presentation of the course of the terrain enables precise analysis of the course of the area and identification of potential places of danger. Based on a contour map analysis or area analysis, endangered areas are used. The number of terrain profiles depends on the complexity of the area, its size, type of facility and the investor's order. Profiles represent characteristic vertical planes.

Generating a profile in the Surfer program can be carried out:

1) automatically - indication of the course of the profile is carried out by marking the cursor along the planned course. The disadvantage of the method is the inability to reproduce the profile with the same course.
2) Manually - creating a profile requires the use of digitizing function in the place of the planned cutting plane and using the slicing function (Slice) to create a profile on a given plane. The method allows free playback of the section or creating the same place on different models.

Each of the stages of reclamation work requires monitoring, in particular determination of settlements of works carried out and stability of designed slopes. Due to the characteristics of the area, the required accuracy and the speed of obtained observations, we can use leveling (geometric or trigonometric) or the GNSS RTK or RTN method.

The reclamation project should take into account the impact of changes in water relations in the area of study. This applies in particular to the excavations, where at the extraction stage as part of the conducted works there may have been a reduction in the free water level and partial subsidence of the area around the excavation. Restoring areas of utility values may result in the restoration of the original water level and secondary uplift. The creation of a depression funnel at the extraction stage may result in impacts in the area several kilometers from the place of work in the form of subsidence. On the other hand, the restoration of water relations may cause uplifting, threatening the safety of objects on the area of impact.

5 EVALUATION OF METHODS OF GEODETIC ANALYZES IN RECLAIMED AREAS

The choice of surface modeling method depends on its shape and distribution of measurement pickets. Interpolation methods based on different algorithms generate different model accuracy. The denser the grid, the more accurate the model will be. Simple models based on linear interpolation give the best results in regular, systematic rectangular, hexagonal, hexagonal networks. Advanced methods (eg kriging) allow extrapolation outside the research area and do not require a constant distance between the measurement points and the regularity of their distribution. This ensures the convenience of measuring and allows you to reduce the cost of measuring work.

The volume of the analyzed area is determined by means of approximate methods. The accuracy of the determined value will increase with the increase of the density of the measurement points, i.e. the size of the elementary mesh of the GRID mesh will be smaller.

The analysis of terrain is one of the basic points in the process of determining the direction of reclamation. The assessment of the slope angle allows the reclamation designer to analyze the stability of the slopes, the possibility of ground runoff and the assessment of the direction of runoff of surface waters. The examined methods of data interpolation have slight differences in the angle of inclination. The RBF interpolation method shows larger maximum values of the slope from the kriging method. The median for both algorithms is the same regardless of the measurement method.

6 CONCLUSIONS

Modern measurement methods provide the ability to quickly measure dense point clouds that reflect the actual terrain. However, they require expensive equipment and specialized software as well as computer units with high computing power. Post-processing of modern methods becomes tedious and, in addition, does not always succeed. The alternative is the classic measurement methods, which include, for example, tacheometry and GNSS. The duration of the measurement with classical methods is longer, but the development of the results is much easier, only requires the use of appropriate calculation algorithms and data modeling.

The performed analyzes confirmed the effectiveness of the method of radial base functions and kriging using an irregular measurement grid due to the inhomogeneity of the algorithms used. In the case of a smaller number of points, the kriging algorithm is more effective, with more data the models give very similar graphical results.

The use of the RFB or K algorithm in the case of examination of the terrain surface is the most efficient. Generated surfaces run closest to the measurement data with realistic and reliable surface mapping. Due to the use of modern measurement methods in the form of laser scanners or drone, additional analyzes should be conducted in order to be able to compare measurements from classical methods with point clouds.

Statistical analyzes of the surface model: volume, inclination angle and terrain profiles support the designer in the preparation of project documentation. The reliability of the results depends directly on the density of the measurement network and the choice of the interpolation algorithm.

REFERENCES

Guidelines of the National Fund for Environmental Protection and Water Management: Guidelines on requirements for reclamation processes, including macro-leveling, carried out with the use of waste, 2009.

Blistan P., Kovanic L., *Aplication of Selected Geodetic Methods in Opencast Quary with Aim of Their 3D Model*, in the journal: Inżynieria Mineralna, Volume: R.15, Journal 1 (2014), pp:287–298.

Cressie, N.A.C.: *Statistics for spatial data. Revised edition*, Wiley & Sons, New York 1993.

Davis J.C.: *Statistics and Data Analysis in Geology*, Wiley & Sons, New York,1986.

Frąckiewicz P., Gil M., Wilk K.; *GIS modeling – searching for optimal location of observation network points in open – pit mining*, conference materials: Sustainability - Environment - Safety 2018.

Hengl, T., Minasny, B., Gould, M.:*A geostatistical analysis of geostatistics*, 2009. Scientometrics, vol. 80, 2009, pp. 491–514.

Namysłowska-Wilczyńska B.: *Geostatystyka. Teoria i zastosowania*. Oficyna Wydawnicza Politechniki Wrocławskiej, Wrocław 2006.

Surfer - User's Guide, Golden Software, 2002.

Zawadzki, J.: *Metody geostatystyczne dla kierunków przyrodniczych i technicznych*, Oficyna Wydawnicza Politechniki Warszawskiej, Warszawa 2015.

Zawadzki J., Cieszewski C. J., Zasada M.,, Lowe R.C.: *Applying geostatistics for investigations of forest ecosystems using remote sensing imagery*, Silva Fennica 2005, vol. 39, pp. 599–617.

Zhou, F., Guo, H.-C., Ho, Y.-S., Wu, C.-Z.: *Scientometric analysis of geostatistics using multivariate methods. Scientometrics*, vol. 73, 2007, pp. 265–279.

Advances and Trends in Geodesy, Cartography and Geoinformatics II –
Molčíková, Hurčíková, & Blišťan (eds)
© 2020 Taylor & Francis Group, London, ISBN 978-0-367-34651-5

Example of combined airborne and terrestrial LiDAR for purposes of engineering-geological analyses

M. Fraštia & M. Marčiš
Slovak University of Technology in Bratislava, Slovakia

M. Bednárik & R. Holzer
Comenius University in Bratislava, Slovakia

ABSTRACT: The paper presents a combined terrestrial and airborne laser scanning, which provided in the difficult terrain sufficient data for the initial engineering-geological assessment of the state and risks of the rock body, endangering the expressway R2 between the towns of Budča and Zvolen, Slovakia. Considering the character of the site – steep, hardly accessible walls in the dense forestation, the laser scanning technology was chosen for its contactless documentation. Results served as basic material for engineering-geological (IG) evaluation, when airborne laser scanning (ALS) provided mainly geomorphological characteristics of interest area, while terrestrial scanning (TLS) documented rock walls in high detail to evaluate potential risk of blocks collapse.

1 INTRODUCTION

Landslides and rock collapses are a very common phenomenon in our geographical conditions that directly endangers human society, its infrastructure and property. According to the State Geological Institute of Dionýz Štúr slope deformations occupy 5.25 % of the total area of the Slovak Republic (SGIDŠ 2018a). The total number of slope deformations documented within the Atlas of slope stability maps of the SR at a scale of 1: 50,000 is 21,029; of which 4794 are small-scale and 16,235 are large-scale (SGIDŠ 2018b).

Engineering-geological (IG) evaluation of steep walls of unstable rock massifs requires, besides knowledge of engineering geology, also other data such as hydrometeorological conditions, geomorphology of the massif and its surroundings, mapping of discontinuities, measurement and analysis of their orientation (direction of azimuth and inclination), volume and mass characteristics. An important role is played by a personal on-site visual inspection by engineering geology experts, direct measurement using a geological compass in a rock massif (Pavlík 1981). In the cases when rocky slopes are inaccessible, it is necessary to hire a climbing team for such measurements, usually trained for that purpose. We know from experiences that it is often not possible to obtain the required data this way. The reason could be difficult accessibility and risk of unstable sites, difficult orientation in morphologically exposed terrain, or simple, the given staff is not able to fulfil these tasks. Contactless measurement technologies with high level of detail, such as scanning methods, are able to very effectively reach such hard accessible locations. Two basic scanning technologies are currently used for these cases - laser scanning (Vosselman, G. & Maas, H. G. 2010) and image scanning (based on structure from motion method) (Westoby, M.J. et al. 2012). Terrestrial laser scanning (TLS) is commonly used for documentation of unstable road cuts (Fraštia, M. 2012), rocks (Wagner, P. et al. 2010) or landslides (Fraštia et al. 2014), but mostly without vegetation cover. Airborne laser scanning (ALS) is used effectively for geo-relief documentation in large and hard to reach areas, when the lower level of detail (1-50 points/m2) is sufficient (Bobáľ et al. 2015).

2 LOCALITY AND PROBLEMS

The site is situated above the expressway R2 between the village of Budča and the town of Zvolen in the station about 232.40 km – 232.90 km. The rock massif is a complex of rock towers, which are located in densely overgrown beech forest, in open areas mostly covered with bushes and grass. The site itself is approximately 500 m long along the expressway and the elevation of the terrain above the road to the top of the rocks and the ridge is up to 150 m (300 - 450 m. a. s. l.). The rock towers are rugged, even tens of meters high and steep, reaching 90° and often overlapping (Figure 1 below), which prevents the possibility of measuring from all sides. The locality belongs orographically to the Štiavnické vrchy mountains and the rock consists mainly of volcanic epiclastics.

In the mentioned road segment, a rock block of dimensions of 2.0 x 1.8 x 1.2 m^3 (approx. 4.3 m^3) collapsed on June 28, 2017 and overcame the retaining barrier built above the road. Some smaller boulders rebounded from both road lanes, broke through the fence and stopped only in the immediate vicinity of the railroad. During the operation of the road, it was already the second recorded case of rocks falling on the road. The first one occurred in 2007 and resulted in the construction of a catchment barrier over the redeveloped cut-off, which proved to be ineffective and was overcome by the last collapse of the rock block (Šimeková, J. 2017). Subsequently, the damaged lane was closed to the public. The specificity of the documentation of this site lies mainly in the problematic terrain: difficult accessibility due to steep slopes and nearly vertical rock cliffs, large elevation, dense forestation and high segmentation of rock blocks and towers.

Figure 1.   Location of the site within the Slovakia (top left), top view (top right), bird's perspective (bottom).

3 DATA FIELD COLLECTION

The primary objective of airborne and terrestrial laser scanning was to create a planimetric and elevation map of the site to plot the locations of individual blocks or groups of unstable blocks, profiles and, in particular, to create the analyse of the risk of falling blocks from the rock massif.

3.1 *Engineering-geological survey*

The on-site IG survey in August 2017 visually documented and evaluated 33 risk blocks with an estimated size ranging from a few m^3 to several tens of m^3 (Šimeková, J. 2017). The movement of geologists on the ground was problematic and often it was unable to get to selected blocks.

3.2 *Airborne laser scanning*

The airborne laser scanning system Leica ALS 70 (Leica Geosystems 2018) was used for measurement. The scanning was performed by the firm Photomap, s.r.o. on April 19, 2018 from flight level 1550 m. a. s. l. (1100 m above the ridge of the hill) in 2 flights. The trees were already heavily foliaged. The whole site with significant longitudinal overlap from the required range contains about 34 million points from ALS including vegetation (Figure 2). Accuracy of point clouds is according to scanner technical specification 0.15 m in position and 0.07 m in height. The density of points in uncovered areas without vegetation reached up to 50 points/m^2, a scanning step of up to 0.14 m. In the forest environment, the density of points on the ground reached about 10 points/m^2 (step 0.30 m) due to significantly dense vegetation.

3.3 *Terrestrial laser scanning*

The terrestrial laser scanner RIEGL VZ400 (Riegl 2018) with a measuring range up to 600 m was used for ground scanning. Scanning was carried out from 13 stations, of which 9 directly from the R2 road and 4 from the positions between railway and Hron river (Figure 4). The spatial accuracy of an individual cloud point relative to the scanning station is 6 mm according to the scanner's producer technical specification. In total about 250 million points were scanned. The average geometric resolution (spacing of points) on vertical rock walls without vegetation reached 0.03 m, which represented up to 1000 points/m^2.

3.4 *Reference network*

Matching of individual scans was ensured by means of 10 targets signalised by a reflection foil with a diameter of 100 mm. Due to the large object distances (200 - 350 m), the targets were self-made and placed in the terrain in the upper parts of the rock massif.

Figure 2.   Axonometric view of the resulting aerial scan (34,000,000 points).

The targets were measured by the Trimble S8 Total Station (Trimble 2019a) during scanning operations with a 10 mm spatial accuracy relative to the local network. At least 4 targets were automatically detected and scanned in each laser scan. The TS positions were also measured by the GNSS apparatus Trimble R8 GNSS (Trimble 2019b) in the system of Slovak Permanent Observation Service (SKPOS, GKÚ 2019) with an accuracy of 0.02 m in position and 0.05 m in height. The local geodetic network formed by the TS stations was subsequently transformed by the Helmert transformation with scale equals 1 to the coordinates determined by the GNSS measurement. Thus the inaccuracy of the SKPOS system was not transferred to the local geodetic network. The output coordinate system is national system S-JTSK and height system Bpv (Baltic after adjustment).

4 DATA PROCESSING

The ALS point clouds from 2 flights were first joined and then automatically classified into the "ground" class. CloudCompare software (EDF R&D 2019) and its CSF filter module were used for that classification. However, due to the complicated geomorphology of the site, a number of points were poorly classified and some rock cliffs themselves were classified as vegetation (Figure 3 left). For this reason, a lot of manual work had to be done to correctly classify the points. The resulting cloud of relief points of interest area from airborne scanning contained 6,500,000 points.

Raw TLS data were processed in the RiSCAN software (Riegl 2019), where the clouds were oriented through the control points georeferencing procedure. The fitting accuracy is given by residues on the CPs which did not exceed 0.03 m in all three axes Y, X, H of the coordinate system. Each scan was georeferenced to a minimum of 4 CPs. After georeferencing, individual scans were filtered from points above relief (from vegetation). This was done first automatically and then manually, as many points were again incorrectly classified. The entire TLS point cloud after filtration and segmentation into the region of interest contains 165,000,000 points (Figure 3 right). Further processing consisted of:

- merging of point clouds from terrestrial and airborne scanning,
- reduction of the density of points on areas outside the rocks,
- segmenting (trimming) the clouds into the area of interest,
- and other manual cleaning of the combined clouds.

The merging of the ALS and TLS point clouds in the LAS data format was only based on their coordinates, since both datasets were georeferenced in the same coordinate and height system. In addition, the automatic alignment of TLS and ALS clouds was not possible as the scanned terrain did not provide sufficiently geometrically high-quality overlapping areas. By visual inspection and by several profiles through the combined clouds, deviations of ALS and TLS with maximum values up to 0.20 m were found, which is within the supposed accuracy of ALS.

The resulting segmented and merged point cloud (Figure 4) was resampled to 24,000,000 points with the preference to preserve rocky parts for practical reasons such as handling a large dataset in others software.

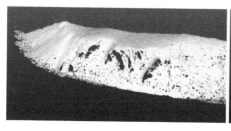 

Figure 3.   Point cloud from ALS after automatic filtering to class „ground" (left) and TLS cloud (right).

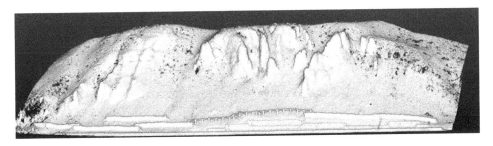

Figure 4.   View on the resulting ALS and TLS merged point cloud.

5   RESULTS

In the resulting point cloud, individual rock blocks were visually identified in cooperation with an expert - geological engineer, which were then segmented and coloured and numbered (Figure 5). This work was carried out in TRIMBLE RealWorks v10.4 software (Trimble 2019). The number of points on individual blocks was in the order of tens of thousands. The volume of the segmented blocks was determined by software. The height of the blocks, their orientation and slope were manually measured directly on the point cloud.

Digital terrain elevation model (DTM) in the raster GEOTIFF form and triangulated digital model in vector TIN form were another outputs. Furthermore, profiles were generated automatically at selected locations through the created DTM (Figure 6).

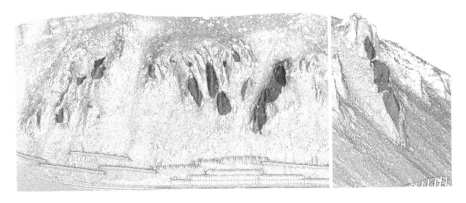

Figure 5.   Identification of rock blocks on a point cloud and their marking and measuring.

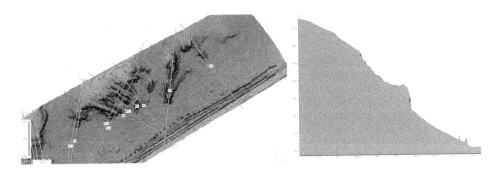

Figure 6.   DTM with profile drawing (left) and example of profile (right).

6 CONCLUSION AND DISCUSSION

The aim of these measurements was to obtain a basis for engineering-geological risk assessment of the site. Creation of a detailed model of the locality in digital form represented by terrain allowed to derive morphological parameters of rock blocks and their surroundings (surrounding slopes) and determine other geometric parameters of these blocks such as volume, height and discontinuities for assessment of consequences of future possible collapse.

Airborne and terrestrial laser scanning provided a comprehensive and accurate macro view of the entire site's morphology, including rock towers and blocks, which could not be obtained from field survey. Remote, non-contact measurement of the physical state of the rock mass by means of the above mentioned methods, supplemented by field survey, provided a much broader view of the area under investigation.

After engineering-geological and geodetic survey and analysis of the locality the possible risk rock blocks and walls were presented and characterized in the unified documentation. Each documented block or slope part was precisely defined according to the "intensity" of instability. Physical field measurements of discontinuities and volumes coincided well with measurements on point cloud, although scanning detail at some locations was not sufficient to detect fissures and vice versa, some blocks documented by scanning were not documented by field survey due to inaccessibility.

Despite the integrated approach - terrestrial and airborne scanning, it was not possible to capture all the rock surfaces and details due to the dense vegetation cover and the very rugged shape and layout of the rock formations. Terrestrial scanning directly within the site was not possible due to terrain conditions (slope inclination up to 40°). Areas recorded only from aerial scanning have an average density of about 40 points/m^2, open areas scanned terrestrially reach densities up to 1000 points/m^2. The density of points from aerial scanning is sufficient to document the morphology of the surrounding terrain, but not for a detailed reconstruction of the geometry of individual rock blocks. Overall the methodology used in this work can be considered as very successful and promising for similar sites, as confirmed mainly by colleagues in engineering geology and geotechnics.

ACKNOWLEDGEMENTS

This article was created with the support of the Ministry of Education, Science, Research and Sport of the Slovak Republic within the Slovak Research and Development Agency, project no. APVV-18-0472.

REFERENCES

SGIDŠ 2018a. *Landslides in Slovakia*. (2018, September 8). Retrieved from: https://www.geology.sk/2018/03/01/zosuvy-na-slovensku/

SGIDŠ 2018b. (2018, September 8). *Atlas of slope stability maps of the Slovak Republic*. Retrieved from: http://apl.geology.sk/mapportal/#/aplikacia/53

Pavlík, J. 1981. Geotechnical methods of rock walls stability determination. Praha: SNTL. p. 215.

Vosselman, G. & Maas, H. G. 2010. *Airborne and terrestrial laser scanning*. Dunbeath: Whittles. 2010. ISBN-13: 978-1439827987.

Westoby, M.J., Brasington, J., Glasser, N.F., Hambrey, M.J., Reynolds, J.M. 2012. Structure-from-Motion photogrammetry: A low-cost, effective tool for geoscience applications. *Geomorphology* 179, 300–314.

Fraštia, M. 2012. Laser versus image scanning of stone massifs. *Mineralia Slovaca*. ŠGÚDŠ: Bratislava, 2012. Vol. 44, No. 2 (2012), ISSN 0369-2086. pp. 177–184.

Wagner, P., Ondrejka, P., Iglárová, Ľ., Fraštia, M. 2010. Current trends in slope movements monitoring. *Mineralia Slovaca*. Vol. 42, No. 2 (2010), ISSN 0369-2086. pp. 229–240.

Fraštia, M., Marčiš, M., Kopecký, M., Liščák, P., Žilka, A. 2014. Complex geodetic and photogrammetric monitoring of the Kraľovany rock slide. *Journal of Sustainable Mining*. Vol. 13, no. 4 (2014). ISSN 2300-1364. pp. 12–16.

Bobáľ, P., Fraštia, M., Ivaňák, M. 2015. Point cloud as data source for DTM creation. *GIS Ostrava 2015. Současné výzvy geoinformatiky*. Ostrava, ČR, 26. - 28. 1. 2015. 1. vyd. Ostrava: VŠB, 2015, online. ISSN 1213-239X. ISBN 978-80-248-3678-2.

Šimeková J. 2017. Emergency landslide on the R2 Budča-Zvolen expressway km 232.565 – 232.900. *Orientačný inžinierskogeologický prieskum*. Žilina: Geotrend. p. 21.

Leica Geosystems 2018. *ALS70-CM*. (2018, September 8). Retrieved from: https://hds.leica-geosystems.com/en/Leica-ALS70_94516.htm

Riegl 2018. *VZ 400*. (2018, September 8). Retrieved from: http://www.riegl.com/uploads/tx_pxpriegldownloads/10_DataSheet_VZ-400_2017-06-14.pdf

Trimble 2019a. *S8*. (2019, March 7). Retrieved from: http://trl.trimble.com/docushare/dsweb/Get/Document-390412/022543-410H_TrimbleS8_DS_0115_LR_sec.pdf

Trimble 2019b. *R8 GNSS*. (2019, March 7). Retrieved from: https://geospatial.trimble.com/products-and-solutions/trimble-r8s

GKÚ 2019. *SKPOS*. (2019, March 7). Retrieved from: http://www.skpos.gku.sk/

EDF R&D 2019. *CloudCompare*. (2019, March 7). Retrieved from: https://www.danielgm.net/cc/

Riegl 2019. *RiSCAN Pro*. (2019, March 7). Retrieved from: http://www.riegl.com/index.php?id=221

Trimble 2019. *RealWorks v10.4*. (2019, March 7). Retrieved from: https://geospatial.trimble.com/products-and-solutions/trimble-realworks

Advances and Trends in Geodesy, Cartography and Geoinformatics II –
Molčíková, Hurčíková, & Blišťan (eds)
© 2020 Taylor & Francis Group, London, ISBN 978-0-367-34651-5

Documentation of old surveying instruments

P. Hánek, P. Hánek & K. Vacková
Topography and Cartography, Research institution of Geodesy, Zdiby, Czech Republic

ABSTRACT: The purpose of this article is to acquaint the readers with the activities realized during the project Surveying and astronomical instruments used in the Czech lands from 16th to 20th century. The aim of the project is to develop methodologies and procedures that enable the preservation and unified information management about old surveying and astronomical-geodetic instruments. During the theoretical part of the processing, the descriptive in-formation about instruments stored in a number of libraries, archives and depositories is collected. The technical solution is divided into two parts. The first part the technical parameters of the instruments including their accuracy specifications are determined. The second part is devoted to acquisition of photographic documentation of instruments and creation of virtual 3D models. Part of the solution is also to create software. This software will allow the management of the collected data by the owners of instruments and their presentation in multimedia kiosks.

1   INTRODUCTION

The aim of the project is to create procedures and methodologies that will enable unified information management about old surveying and astronomical-geodetic instruments. For the purposes of record keeping is developed information system of old surveying instruments. Created solution also enable to use and publish information about the instruments in digital form on the web portal or in the interactive information panels in museums. The results of the project will be presented at the exhibition of surveying and astronomical-geodetic instruments. This exhibition is scheduled for 2021 and will take place at the National Technical Museum (NTM), which is partner in the project.

2   INFORMATION SYSTEM OF OLD SURVEYING INSTRUMENTS

Information system consists of two parts, administrator and user. The administrative section will serve the museum staff for entering and editing data about registered instruments. The user section will make the collected information available to the general public. At present, the work is focused on creating the administrative part and collecting information about selected instruments. When selecting instruments, the instruments manufactured in Czech workshops and companies, or in neighbouring Europe, will be preferred.

2.1   *Technical solution of information system*

As part of the administration part, a data model had to be designed. This data model contains all information collected about the instruments, including the necessary additional information. The main categories of collected information are categorization and description, art photography, technical photography, 3D model, technical parameters and iconographic information. Individual categories are described in the following chapters of the article. Additional information includes registration numbers, owners, administrator contacts, etc. A simplified version of the data model is shown in Figure 1. The information system is created using the open source solution. As operating system was chosen Linux Ubuntu, Apache Web server and SQL datatabase MySQL. This combination can be replaced with other software in the future. As a scripting language is used PHP, Java and Javascript.

2.2 Data collection for the information system

2.2.1 The main categories of collected information

2.2.1.1 CATEGORIZATION AND DESCRIPTION OF INSTRUMENTS
Currently, surveying instruments are divided into 26 categories within the project solution, which are further divided into subcategories. The main categories include for example a levelling instrument, theodolite, etc. Part of categorization is to determine the manufacturer, model and production date of the instrument.

The instruments description is a short annotation. This annotation contains primarily a short summary description, information about history and description of construction and other curiosities. Within this part of the solution, the devices are interconnected with important personalities (designers, surveyor) and the activities in which they were used.

2.2.1.2 ART PHOTOGRAPHY
Art photography is used for the purpose of publishing high-quality images of surveying equipment in posters at exhibitions, exhibition catalogues and printed or electronic publications. Photographs are taken by photographers of the National Technical Museum in a professional studio. A sample photo is shown in Figure 2.

Figure 1.   Simplified data model schema (VÚGTK, 2019).

Figure 2.   Art Photography - theodolite Gerb. Fromne (VÚGTK, 2019).

2.2.1.3 TECHNICAL PARAMETERS

Technical parameters of the instruments are determined in order to be able to compare the instruments with each other. Knowledge of the technical parameters is also important for disciplines dealing with history and thanks to the knowledge of instrument accuracy, it is possible, for example, to specify information about the accuracy of cartographic products.

The following information are always collected:

- dimensions - length, width, height,
- weight,
- existence of original transport container,
- image of storage in a transport container.

In addition are determined parameters which indicate the accuracy of the instrument and the quality of its production. These parameters are specific to each instrument category respectively individual instruments.

These include:

- telescope magnification,
- used scales (reading tools),
- division of scales,
- scale resolution,
- used units,
- measurement accuracy,
- level sensitivity.

These parameters are determined using elementary measuring tool, but also using the know-how of the Certified Calibration Laboratory of Research Institute of Geodesy, Topography and

Cartography. Especially during the determination of the measurement accuracy, adapted laboratory calibration procedures of Calibration Laboratory and procedures specified in ISO 17123 are used.

2.2.1.4 TECHNICAL PHOTOGRAPHS AND 3D MODEL

Technical photographs and 3D models are made by Research Institute of Geodesy, Topography and Cartography staff in their own photographic laboratory. Technical photographs capture interesting details of the instrument. It is a detail of a reading tool or a significant detail of the structural design of the instrument.

3D models are created in order to present the instruments to a greater extent at exhibitions not focused on surveying, but on related fields. Virtual models of instruments will be able to be presented to the public at the information kiosks.

Taking technical photos and images used for the 3D model is done with the Canon Mark IV 5D camera, which is equipped with three Canon USM prime lenses (14, 24 and 50 mm) and one Canon macro lens.

For the purposes of creating 3D model was developed a new special device which is shown in Figure 3. The device is equipped with a stepper motor to ensure the automatic rotation of the table in specified steps. The device engine and camera shutter are connected to the control unit which can be managed from a computer. In the control software, it is possible to specify the number of images per rotation, the maximum number of images is 360. For our needs, we use the 24-72 images per rotation. The number of acquired images depends on the design of the instrument. Each instrument is captured in 3 - 4 rows in vertical position and in 2 rows in horizontal position. For subsequent image processing is used graphics software Affinity Photo, where individual images are masked. The final 3D model of the instrument is created in photogrammetric program Agisoft Metashape. An example of creating a 3D model is in Figure 4.

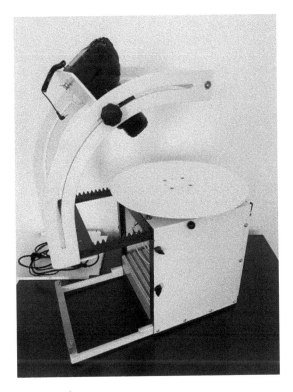

Figure 3.   Device for digitizing (VÚGTK, 2018).

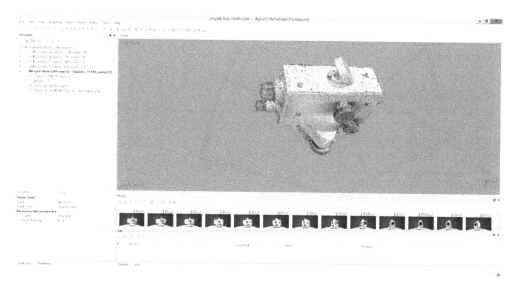

Figure 4. Screenshot Agisoft Metashape (VÚGTK, 2019).

2.2.1.5 ICONOGRAPHIC MATERIALS

In this part of the project, the extant historical written and pictorial materials are interconnected with the instruments. These materials are mainly brochures, production documentation, user manuals, historical photographs and, in rare cases, measurement records. Some of these materials are also converted into digital form by employees of the National Technical Museum.

2.2.2 *Secondary collected information*

A selection of basic cataloguing information that is maintained for each instrument by its owner. For information system management purposes, these parameters are extended with information about the owner, including contact details of his representative.

3   COLLECTIONS

During the realization of the project, cooperation was established with institutions and a number of collectors who have the most significant collections of geodetic and astro-geodetic instruments in the Czech Republic. Such cooperation enables us to select representative samples of instruments to fill the information system. However, it is necessary to indicate the collection of our partner (The National Technical Museum) as the most important collection. This collection contains approximately 1800 surveying and 500 astronomical exhibits. Part of this collection are also unique instruments in the world from the time of Emperor Rudolf II. and the production of important Czech workshops Brandejs, Spitra, Koula, Haase & Wilhelm, J. and J. Frič, Srb and Štys, Meopta and others. In terms of historical and cultural value, this collection can be compared with the world's leading collections.

4   CONCLUSION

The result of the project should be an information system for the unified management of information about surveying instruments stored in various collections and archives. Created virtual models should be useful for the presentation of instruments at exhibitions in digital information kiosks. Last but not least, the instrument documented during the project will be available on the website http://www.surveyinginstruments.org. This web portal will allow the general public to use a new source of information about an important group of national cultural heritage.

This contribution was created as a part of project no. DG18P02OVV054 „Surveying and astronomical instruments used in the Czech lands from 16th to 20th century" under the auspices of the Program for applied research and development of national and cultural identity (NAKI) of Ministry of Culture of the Czech Republic.

Advances and Trends in Geodesy, Cartography and Geoinformatics II –
Molčíková, Hurčíková, & Blišťan (eds)
© 2020 Taylor & Francis Group, London, ISBN 978-0-367-34651-5

Static load test measurement of reinforced concrete columns using terrestrial laser scanning

R. Honti & J. Erdélyi
Department of Surveying, Faculty of Civil Engineering, Slovak University of Technology in Bratislava, Bratislava, Slovakia

J. Dobrý
Department of Concrete Structures and Bridges, Faculty of Civil Engineering, Slovak University of Technology in Bratislava, Bratislava, Slovakia

ABSTRACT: The technology of terrestrial laser scanning (TLS) has become increasingly popular in deformation monitoring of buildings or building structures. The use of TLS during static load test of reinforced concrete columns is becoming a common practice all over the world. The advantages of TLS are the efficiency of spatial data acquisition and the contactless determination of the 3D coordinates of points lying on the surface of the scanned object. The next advantage is the scan rate, which reaches 2 million points/s with current scanners, what allows significant reduction of time needed for the measurement, respectively increase of the quantity of the acquired information about the surveyed object. The result of the measurement is a point cloud, which covers the whole surface with 3D points of the measured objects, which is also a big advantage over conventional contact measurement techniques used in load tests, which can measure only one single point at a time. This means that by processing of point clouds it is possible to monitor the deformation on the whole surface of the object measured.

The paper deals with the static load test of 4 columns tested in the laboratory of TU Wien. The measurements were accomplished with Trimble® TX5 3D laser scanner. The measured objects were reinforced concrete columns with dimensions of 5 m (length) x 0.15 m (height) x 0.35 m (width).

The object measured, the measurement procedure, the data processing and the results of the loading tests are described. The results obtained by TLS were verified with conventional contact measurement techniques during the load tests.

1 INTRODUCTION

Building structures (e.g. reinforced concrete columns) during static load test are mostly monitored using conventional contact measurement techniques like linear variable differential transformer (LVDT) units, strain gauges and deformeters. These conventional measurement techniques deliver high degree of accuracy of the changes in the length measured, but the limitation is as they offer only 1D measurements between points and the data can be gathered only for specific points (deformation determination in predefined discrete points). So, in some cases they might be not effective for testing larger complex objects or for the deformation analysis of the whole structure tested. Nowadays we are seeing a significant increase of using other techniques for this purpose, e.g. the technology of terrestrial laser scanning (TLS) (Schäfer et al., 2004), (Yang et al.,2017), (Zogg et al., 2008), (Erdélyi et al., 2018), (Monserrat et al., 2007), (Abellán et al., 2009), (Mukupa et al., 2017) digital photogrammetry (Marčiš et al., 2017), (Wagner et al., 2017), (Urban et al., 2015), (Hamouz et al., 2014).

TLS allows non-contact documentation of the behavior of the monitored building structure. The scan rate of current scanners reaches 2 million points per second, what allows significant reduction of time necessary for the measurement. The precision can be increased using suitable data processing, when valid assumptions about the scanned surface are available (Vosselman et al., 2010). In many cases it can be solved by fitting geometric shapes to the point cloud to approximate the chosen parts of the object measured. The result of the TLS measurement is a set of 3D point (point cloud) covering the whole surface of the object monitored.

The motivation of using TLS is to gather data from the entire surface of the structure monitored without the need of use some specific deformation monitoring targets. Using suitable data processing of point clouds TLS offers 3D deformation monitoring of the whole surface of the object measured, not only 1D monitoring of chosen points like with the conventional measurement techniques.

The paper briefly introduces the experimental static load test measurement of reinforced concrete columns performed using TLS, compared to other conventional contact measurement techniques.

2 CHARACTERISTICS OF THE STATIC LOAD TEST

The aim of the TLS measurements was the static load tests performed on 4 slender compressed reinforced concrete columns. The columns were made with the following parameters:

Material: Concrete C50/60;
Reinforcement: B 500B;
Steel: S 355;
Dimensions: 5 m (length) x 0.15m (height) x 0.35m (width) (Figure 1);
Weight: 670.28 kg.

On the both ends of the columns, steel plates of thickness of 25mm were placed and all the longitudinal reinforcement was welded to them.

During the experiment, the columns were loaded with hydraulic press situated on the ends of the columns (Figure 2). The loading process was performed in 4 and 5 steps in the case of the first 2 columns and in 9 steps in the case of the last 2 columns (Table 1). Between each load interval, there was a non-load period of 10-15 min, when the TLS measurements were performed.

In the Figure 2 a simplified representation of the testing rig can be seen with all the marked measuring units. For the verification of the force, 3 load cells were placed into the test rig. The center of the gravity of these load cells was in the axis of the column. The force was also checked by the pressure of the oil in the test rig. The presses were also able to measure their

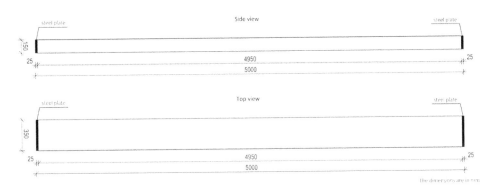

Figure 1.   The shape of the reinforced concrete columns from side view (up), top view (down).

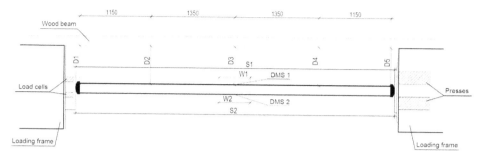

Figure 2.   Simplified representation of the test rig.

Table 1.   The load stages of the testing.

| Epoch | Load Stage [kN] | | | |
	Column 1	Column 2	Column 3	Column 4
1	111	94	103	106
2	589	396	394	398
3	735	541	588	587
4	846	684	643	641
5	-	804	694	693
6	-	-	735	736
7	-	-	785	765
8	-	-	820	795
9	-	-	820	817

Figure 3.   The reinforced concrete column tested with the measuring units installed.

extrusion. In addition to these parameters, several measuring units were installed directly on the concrete column, which measured its deformation (Figure 3).
The measuring units mentioned were as follows:

• D1-D5 are LVDT measuring units used for the longitudinal deformation of the concrete column, the accuracy of these units is ±0.1% of the full-scale reading, what represents 0.4 mm in this case;
• DMS1 and DMS 2 are strain gauges used to measure the relative deformation of the top and the bottom surfaces of the concrete column;
• W1 and W2 are also LVDT units for relative deformation measurement on the top and the bottom surfaces of the column. They were placed in the middle of the column with span of 50cm;
• S1 and S2 are measuring units for the column's compression and check if the adapt steel plates are not tilting.

In addition, the deformation of the reinforced concrete columns during the static load test was measured by the technology of TLS, this part of the load test is described in the next chapter.

3 LOAD TEST MEASUREMENT USING TLS

The measurements were performed with Trimble® TX5 3D laser scanner (Figure 4 – blue arrow). The reference targets for the TLS were spheres and flat (checkerboard) targets situated evenly in the whole surroundings of the tested structures. These targets created a reference network and were used for the stability check of the instrument on the station (scan position), and for possible point cloud registration. Eventually the registration was not necessary, because the position of the scanner in several epochs did not change (determined by the coordinate differences of the network of the reference points between each epoch).

The scanning was performed in several epochs with the stated loading stages in the Table 1. The measurement was focused on the determination of the static vertical displacements of the reinforced concrete columns. The top and the side surface of the columns were scanned in every non-load period between loading steps from a single position of the scanner (Figure 4). The maximum distance between the scanner and the columns tested was below 5 m and the measurements of a single point were repeated 3 times (repeated distance measurement). With the mentioned scanner and considering the measurement conditions stated, the accuracy of a single measured point was less than 1.2 mm in any cases.

The main task of the data processing was to determine the deformation of the columns measured. The deformation was calculated by comparing point clouds of the tested column from several epochs.

The point cloud processing was performed using CloudCompare software. At first the points lying on the column surface were segmented from the scanned point clouds. Next, the point clouds from the separate loading steps were compared to the reference point cloud, which was the point cloud measured in the first (approximately 100kN) load stage. In the last step the deformations were calculated as nearest neighbor distances. It means that for each point of the compared cloud the nearest point in the reference cloud is found and the distance between these 2 points are calculated.

On the top of Figure 5 the reference point cloud of the second column tested is showed, in the middle the reference cloud and the cloud after the last (5th) loading stage (bottom) are showed. The differences between the 2 clouds are visualized by color, on the left side the color scale is showed, where the values are in meters. The maximum displacement in the middle part

Figure 4. The experiment set up, position of the TLS (blue arrow) and the concrete column tested (red arrow).

Figure 5.   Comparison of point clouds for the 2nd column tested.

Table 2.   Comparison of the result from LVDT units and the TLS measurement for column 2.

Ep.	Load Stage [kN]	D1 [mm]	TLS1 [mm]	D2 [mm]	TLS2 [mm]	D3 [mm]	TLS3 [mm]	D4 [mm]	TLS4 [mm]	D5 [mm]	TLS5 [mm]
1	94	0.0	0.0	0.0	0.0	0.0	0.0	0.0	0.0	0.0	0.0
2	396	0.2	1.0	3.3	3.7	5.2	6.3	3.5	3.3	0.3	0.9
3	541	0.2	1.1	5.6	5.9	8.7	8.7	5.8	6.2	0.4	1.2
4	684	0.1	1.0	9.9	10.6	15.1	14.3	9.9	9.8	0.7	1.4
5	804	0.7	1.5	17.5	18.5	27.0	28.1	17.3	18.4	1.2	1.8

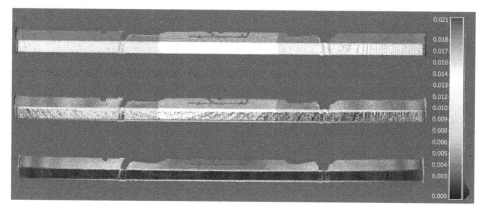

Figure 6.   Comparison of point clouds for the 4th column tested.

of the column was 28.1 mm, and on the ends 1.8 mm. The values of displacements compared to the values from LVDT units are shown in Table 2.

The Figure 6 shows the comparison between the point clouds of the last (4th) column tested from the first and the last (9th) loading stage. The displacement in the middle part of the column was 19.1 mm, and on the ends 1.1 mm. The values of displacement with the values from the LVDT units are shown in Table 3. The maximum value of displacement in the middle part for the 1st column was 23.5 mm (on the end 1.6 mm) and for the 3rd column 28.9 mm (on the end 1.7 mm).

Table 3. Comparison of the result from LVDT units and the TLS measurement for column 4.

Ep.	Load Stage [kN]	D1 [mm]	TLS1 [mm]	D2 [mm]	TLS2 [mm]	D3 [mm]	TLS3 [mm]	D4 [mm]	TLS4 [mm]	D5 [mm]	TLS5 [mm]
1	106	0.0	0.0	0.0	0.0	0.0	0.0	0.0	0.0	0.0	0.0
2	398	0.3	0.7	2.5	2.9	4.0	4.5	2.6	3.4	0.2	0.2
3	587	0.4	0.9	5.1	6.1	7.8	8.4	5.1	6.1	0.4	0.6
4	641	0.4	1.0	6.1	6.4	9.4	9.8	6.0	6.2	0.4	0.6
5	693	0.3	0.8	7.2	7.3	11.0	11.0	7.1	7.4	0.5	0.9
6	736	0.2	1.1	8.7	8.5	13.1	13.7	8.4	8.9	0.6	1.1
7	765	0.2	0.8	9.7	9.3	14.8	14.4	9.5	9.5	0.7	1.2
8	795	0.1	0.9	11.2	11.4	17.1	16.6	10.9	11.2	0.8	1.2
9	817	0.0	0.8	12.9	12.5	19.8	19.1	12.5	12.8	0.9	1.1

The results obtained by processing of TLS point clouds describe the displacements better, because it gives a complex picture about the deformation of the whole surface of the structure tested.

In the last step, monitored points were modelled from 5 nearest points (segmented from the point cloud) in the position of LVDT units. Subsequently the vertical displacement of these points was determined in each measurement epoch. The results were used for comparison of the results from TLS with the data from LVDT (for data verification). The result of the comparison is showed in Table 2 and 3. In the Table 2 the comparison for the second column and in Table 3 for the last column is showed. The maximum difference between the two methods is 1.1 mm in both cases, which is less than the accuracy of a single measured point by TLS.

The maximum difference between the two methods in the case of column 1 was 1.0 mm and 1.2 for the 3rd column. These values are also below the mentioned accuracy of the TLS measurement.

4 CONCLUSION

The results from the current experiment indicates that the technology of terrestrial laser scanning is an effective, accurate and suitable method for displacement determination during static load test of building structures. TLS enables contactless monitoring not only of some selected points, but the whole surface of the object tested, so it can provide a complex model of the object deformations. The differences between the results from the LVDT units and from processing of point clouds were less, than the point cloud accuracy in any cases. The use of TLS can lead to reduction of the costs spent on some conventional measurement equipment, like LVDT units, deformeters, strain gauges, etc. It can lead also to reduction of time needed for the load test preparation, because it eliminates the need of montage some contact measuring equipment on the object tested.

ACKNOWLEDGEMENT

This work was supported by project VEGA 1/0506/18 "Development of an Algorithm for Automated Quality Check of Construction Work in BIM Environment", funded by the Scientific Grant Agency of the Ministry of Education, Science, Research and Sport of the Slovak Republic.

REFERENCES

Abellán, A., Jaboyedoff, M., Oppikofer, T., & Vilaplana, J. M. 2009. Detection of millimetric deformation using a terrestrial laser scanner: experiment and application to a rockfall event. In: Natural hazards and earth system sciences. Vol. 9. No. 2. pp. 365–372. DOI: 10.5194/nhess-9-365-2009

Erdélyi J., Kopáčik, A. & Kyrinovič, P. 2018. Multi-sensor monitoring of suspended steel bridge structure. In: Advances and Trends in Geodesy, Cartography and Geoinformatics - Proceedings of the 10th International Scientific and Professional Conference on Geodesy, Cartography and Geoinformatics, 2017. pp. 27–33.

Hamouz, J., Braun, J., Urban, R., Štroner, M., & Vráblík, L. 2014. Monitoring and evaluation of prestressed concrete element using photogrammetric methods. In: International Multidisciplinary Scientific GeoConference Surveying Geology and Mining Ecology Management, SGEM, Vol. 3(2), pp. 231–238. DOI: 10.5593/SGEM2014/B23/S10.029

Schäfer, T., Weber, T., Kyrinovič, P. & Zámečníková, M. 2004. Deformation measurement using Terrestrial Laser Scanning at the Hydropower Station of Gabčíkovo. In. INGEO 2004 and FIG Regional Central and Eastern European Conference on Engineering Surveying. Bratislava: KGDE Svf STU, 2004, 10 p. ISBN 87-90907-34-05.

Marčiš, M., Fraštia, M. & Augustín, T. 2017. Measurement of Flat Slab Deformations by the Multi-Image Photogrammetry Method. In: Slovak Journal of Civil Engineering. Vol. 25. No. 4. pp. 19–25. DOI: 10.1515/sjce-2017-0019

Monserrat, O., Crosetto, M. 2007. Deformation measurement using terrestrial laser scanning data and least squares 3D surface matching. In: ISPRS Journal of Photogrammetry & Remote Sensing. Vol. 63. No. 1. pp. 142–154. DOI: 10.1016/j.isprsjprs.2007.07.008

Mukupa, W., Roberts, G. W., Hancock, C. M., Al-Manasir, K. 2017. A review of the use of terrestrial laser scanning application for change detection and deformation monitoring of structures. In: Survey Review. Vol. 49. No. 353. pp. 99–116. DOI: 10.1080/00396265.2015.1133039

Urban, R., Braun, J., & Štroner, M. 2015. Precise deformation measurement of prestressed concrete beam during a strain test using the combination of intersection photogrammetry and micro-network measurement. In: Proceedings of SPIE - the International Society for Optical Engineering. Vol. 9528. DOI: 10.1117/12.2184038

Vosselman, G. & Maas, H.G. 2010. Airborne and Terrestrial Laser Scanning, Dunbeath: Whittles Publishing, 2010. 318 p. ISBN 978-1904445-87-6.

Wagner, A., Wiedemann, W. & Wunderlich, T. 2017. Fusion of Laser Scan and Image Data for Deformation Monitoring – Concept and Perspective. In: Proceedings of INGEO 2017 – 7th International Conference on Engineering Surveying, pp.157–164. ISBN 978-972-49-3200-0.

Yang, H., Omidalizarandi, M., Xu, X. & Neumann, I. 2017. Terrestrial laser scanning technology for deformation monitoring and surface modeling of arch structures. In: Composite Structures. Vol. 169.

Zogg, H.M. & Ingensand, H. 2008. Terrestrial Laser Scanning for Deformation Monitoring – Load Test on the Felsenau Viaduct (CH). In: The ISPRS, Beijing, p. 555–562.

Advances and Trends in Geodesy, Cartography and Geoinformatics II –
Molčíková, Hurčíková, & Blišťan (eds)
© 2020 Taylor & Francis Group, London, ISBN 978-0-367-34651-5

The impact of systematic errors of levels on the results of the system calibration

Pavol Kajánek, Alojz Kopáčik, Ján Ježko & Ján Erdélyi
Department of Surveying, Slovak University of Technology in Bratislava, Bratislava, Slovakia

ABSTRACT: Regular calibration of levels and level rods is important to guarantee the prescribed accuracy for precise height measurement such as displacement and deformation measurements. Requirements for precision levelling staffs are described in DIN 18717 Precision levelling staffs. System calibration is used to calibrate the entire levelling system (level and levelling rod). The principle of the system calibration is the comparison the difference of level's measurements and the difference of laserinterferometer's measurements. Laserinterferometer is used as etalon and represents the reference value of the levelling staff reading. During the system calibration, there are used levels with levelling staffs and therefore the error model determined consists from the error of the level and the error of the levelling staff. The paper deals with the impact of the systematic error of the level on the final correction of the levelling system determined by calibration. Experimental measurements were performed by the vertical comparator, which was developed at the Department of Surveying of the Slovak University of Technology in Bratislava. During the experimental measurements, system calibration was performed with three different levels and two barcode levelling staffs. An analysis of the impact of the systematic error of the level on the final correction was made by comparing the calculated corrections for each levelling system. Since the same levelling staff was used for all measurements, the differences in corrections represent a systematic error of the level. The paper brings the description of the vertical comparator structure and data processing for the calculation of corrections, too.

1 INTRODUCTION

During the calibration of levels and levelling staffs, the criteria for precise levelling staffs are evaluated based on the standard DIN18717. This standard defines the maximum correction of the staff reading, the maximum thermal expansion coefficient of invar and the maximum index error (correction of first staff reading). The calibration can be done (1) for the entire levelling system (Woschitz et al. 2003, Takalo et al. 2004, Vyskočil et al. 2015, Ingensand 1999 & Takalo et al. 2002), namely system calibration or (2) only the levelling staff and then it is called staff calibration (Gassner et al. 2007, Wu et al. 2013 & Maurer et al. 1995). The basic difference between these two calibration procedures is the device, which is used for the staff reading. In the case of system calibration, the staff reading is made by the level and therefore the calculated corrections relate to the entire levelling system. For the separate calibration of the levelling staff, camera and image processing were used to define the staff reading.

The article brings results from system calibration, which are used for analysis of the influence of systematic errors of the level at the final results. Resulting the system calibration, corrections are evaluated by comparing the level measurements with the interferometer measurements (Woschitz et al. 2003). For this purpose, it is necessary to guarantee the precise positioning of staff to the predefined position, the staff motion must be linear, and the staff must be in a vertical position (in specific cases, calibration can be performed for staff in horizontal position too). A vertical comparator is used to ensure these conditions. The structure of the comparator is described in the next part of the article. Next, the article describes the determination of corrections of the levelling system and define the parameters of the equation, which enable the calculation of correction of the measured staff reading.

During the experimental measurements, calibration of six levelling systems (created by combination of three levels and two levelling staff) were performed. To express the systematic error of the level, a difference between corrections of two levelling systems consist from two different levels and the same levelling staff was used. Based on the results of all possible combinations of levelling systems, the impact of the systematic errors of the used levels on the results of the system calibration was analysed.

2   VERTICAL COMPARATOR STRUCTURE

To achieve the required accuracy of calibration, it is necessary to ensure the precise positioning of staff, the vertical position of staff, stable atmospheric conditions during whole calibration and the stability of the level and the levelling staff. The advantage of the vertical comparator is the vertical position of the staff, because it is in the same position during the calibration as well as the measurement.

The comparator frame consists from aluminium profiles (Figure 1). Two rails are mounted on comparator frame as railroad for the staff carriage, which is used for positioning of the staff. The comparator frame is fixed to the laboratory wall using five consoles. The stability of the comparator frame is controlled by two dual-axis inclination sensors (Leica NIVEL 220). The first inclination sensor is located on the base of the comparator frame and the second is located on the comparator frame at level of the stepper motor.

The staff carriage consists from a plate which is mounted on the rails by bearings. On the bottom end of plate is the console with bolt for the staff set-up. Four staff's mounting bracket are placed along the plate for stabilization of the staff in vertical position.

The vertical movement of the staff carriage is realized by a stepper motor, where the movement of the motor shaft is transmitted by a gear wheel to the gear rod located backside of the carriage. The microcontroller communicates with three touch switch sensors, where one

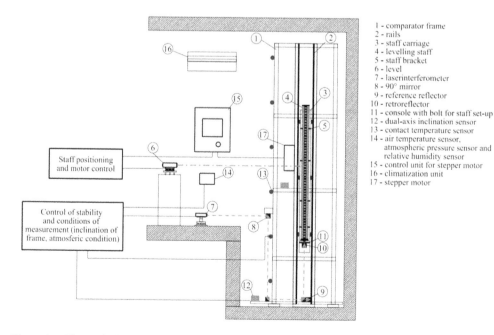

Figure 1.   The vertical comparator scheme.

sensor is used for set the reference position of the staff and two sensors are used as limit switch, which stopped the motor, when the staff carriage reaches the limit position.

During calibration it is necessary to ensure stable atmospheric conditions, especially stable air temperature 20 °C (with a tolerance of ± 1 °C). Therefore, it is necessary to set the laboratory temperature to 20 °C before calibration and then measure the air temperature, the relative humidity and the atmospheric pressure during the calibration.

The level is stabilized on the pillar, which define the length of the levelling sign at 2.86 m. A program for remote control of the level has been created and used, to ensure the instrument stability during calibration.

Positioning of the levelling staff to a predefined position is realized by stepper motor. For the calibration, the length of one step (distance between a two predefined position of staff) was set to 20 mm in both measurement directions. Four measurements are made during the calibration, the staff are passed the whole way twice, back and forth.

3   SYSTEM CALIBRATION

The principle of system calibration is based on the comparison of the level's measurements with the interferometer's measurements (Takalo 2004). The interferometer defines the reference value of the difference between two staff readings (the distance between two staff positions).

The correction of the staff reading v_i is determined from the difference of measurements determined by the levels l_i^L, l_0^L and the measurements from the interferometer l_i^{IF}, l_0^{IF} by formula:

$$v_i = \left(l_i^L - l_0^L \right) - \left(l_i^{IF} - l_0^{IF} \right) \tag{1}$$

During calibration, the staff is set to predefined positions and the corrections are calculated relative to the initial position of the staff. In order to obtain an absolute value of correction (correction relative to the footplate of the levelling staff), it is necessary to determine the correction for the first staff reading, which usually called as index correction. This correction is determined as difference between the reference position of the bar and its measured position. The reference position of the bar is defined from the code pattern of the staff. The measured position of the bar is determined by image processing and interferometer measurement.

The measuring range (0.18 m to 2.94 m) is limited due to the minimum part of the staff required for reading. Based on the length of the calibration step (20 mm) and the measuring range, 139 positions are evaluated in one measurement direction.

The calculated corrections are approximated by a regression line (Figure 2), which parameters are estimated by the least square's method. The slope of the regression line corresponds to a systematic effect of the levelling system, that significantly affects the measurements (staff readings) and therefore needs to be eliminated.

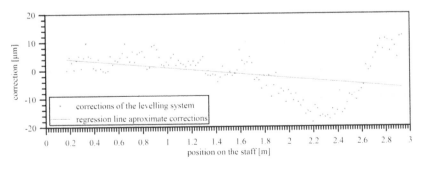

Figure 2.   Correction of levelling system approximated by regression line.

The correct staff reading L could be calculated using the calculated scale of the levelling system m_0, the thermal expansion coefficient of invar α_T and the correction of the first staff reading v_G:

$$L = L' \cdot [1 + m_0 + \alpha_T \cdot (T - T_0) \cdot 10^{-6}] + v_G \qquad (2)$$

where:
$L'[m]$ - measured staff reading,
$T\ [^\circ C]$ - temperature during measurement,
$T_0[^\circ C]$ - temperature during calibration (20 °C).

4 ANALYSIS OF SYSTEMATIC ERRORS OF LEVELS

During experimental measurements, six levelling systems were calibrated. These systems were created by combination of three levels and two levelling staffs. The specification of calibrated levels is in Table 1. The calibrated level differs in software version, the third level has an older software version than the first and the second level.

We can see that levelling system with a third level has bigger scale than scale of the level system in which the first or second level were used (Table 2). Also, the variation of corrections is higher in the case of the third levelling system (Figure 3).

Table 1.   Specification of calibrated levels.

	Level	Serial number	Software	Firmware	Processing technique	Standard deviation of 1km double-run levelling
1st	Trimble DiNi 12	738980	3.40	2.0.1	Edge	0.3 mm
2nd	Trimble DiNi 12	730807	3.40	2.0.1	detection	0.3 mm
3rd	Zeiss DiNi 12	700339	3.31	1.0.1		0.3 mm

Table 2.   Results of system calibration for all levelling systems.

Levelling system	Levelling staff Trimble DL13	Level Trimble DiNi 12	Scale [ppm]	Standard deviation of scale	Standard deviation of corrections [μm]
1st	1st levelling staff (s. n. 62888)	1st level (s. n. 738980)	−3.96	0.77	4.43
2nd		2nd level (s. n. 730807)	−3.67	0.73	5.83
3rd		3rd level (s. n. 700339)	−5.47	0.91	7.89
4th	2nd levelling staff (s. n. 64482)	1st level (s. n. 738980)	−6.39	0.85	4.07
5th		2nd level (s. n. 730807)	−6.09	0.88	4.57
6th		3rd level (s. n. 700339)	−5.88	1.05	10.03

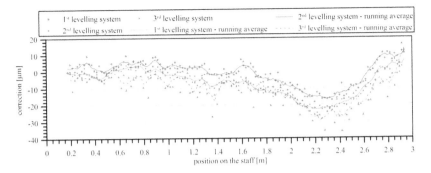

Figure 3. Corrections calculated from system calibration for all levelling system consists first staff.

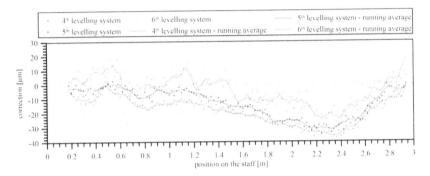

Figure 4. Corrections calculated from system calibration for all levelling system consists second staff.

The analysis of the systematic error of the level is based on the fact that the calculated corrections of the levelling system v_{system} consist from errors of the level v_{level} and errors of the levelling staff v_{staff}. In order to analyse the systematic error of the level, corrections of levelling system were calculated for two different levelling system, which consist from the same staff. The difference between corrections of these levelling systems eliminate the error of the levelling staff. In the results there are only a differences between the systematic errors of the levels.

$$v_{system}^1 - v_{system}^2 = v_{staff}^1 + v_{level}^1 - v_{staff}^1 - v_{level}^2 = v_{level}^1 - v_{level}^2 \qquad (3)$$

Since two levelling staff were used in the calibration, the difference in systematic errors of the two levels were verified on levelling systems, which consisted of the same levels and the second staff. The differences of systematic errors evaluated from these systems were almost the same (Figure 5).

In the same way, the differences in systematic errors were calculated for the other two combination of levels (Figure 6). By comparing the results, we can see that the bigger differences in systematic errors of levels were in the case of a combination where the third level was used. In the case of these combinations, we can see a significant change in the difference for last two staff reading, which was probably caused by the extreme value of the staff reading.

5 CONCLUSIONS

The system calibration of the levelling system allows to calibrate level and levelling staff together. The principle of calibration is the comparation measurements made by level with measurements made by the interferometer. During the calibration, the staff is step by step

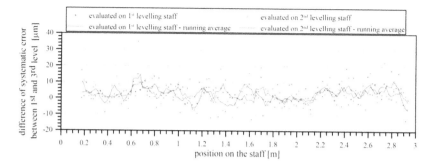

Figure 5.   Differences of systematic errors of two levels express from results of system calibration.

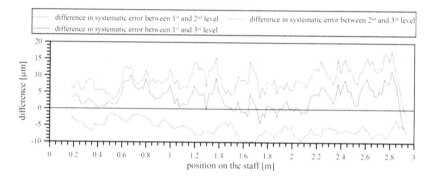

Figure 6.   Difference of systematic errors of two levels express from all combinations.

moved to predefined positions with a step of 20 mm length. The precise positioning of the staff at the predefined positions and their vertical position is provided by the comparator. In our case, the vertical comparator was used, because the levelling staff can be calibrated in vertical position, which corresponds to the real conditions of the measurement.

At the beginning of the article, the construction of the comparator and its main parts, which provided the stability control of the comparator structure and the control of the atmospheric condition, are described.

Results of the system calibration are corrections of staff readings, which consist from systematic error of the level and systematic error of the levelling staff. The main aim of the paper is analysing the impact of the level systematic error on the results of the system calibration. To express this impact, differences between the corrections of two levelling systems, which consist the same levelling staff, were calculated. The difference between the correction of two levelling system eliminate the systematic error of the levelling staff and results consist only the systematic errors of levels.

During the experimental measurements, six levelling systems, which consist of three levels and two levelling staff, were calibrated. Based on the corrections of levelling systems in which the same levelling staff was used, we can see that the scale of the levelling system with a third level is significantly different from the other two levelling systems. For this levelling system, larger corrections were calculated, and the variance of corrections was higher, too. Same results were obtained from the comparison of corrections for levelling systems, which use the second levelling staff.

From the difference between corrections of the levelling systems were evaluated differences of systematic errors of all combination of levels. Based on the comparison of results, bigger differences were determined in the case of the combinations where the third level was used.

The same results were achieved from comparison of the differences between corrections of levelling systems, formed by the second levelling staff.

The results of the system calibration as well as the differences in the systematic errors of the levels show that the third level (s. n. 700339) has a bigger systematic error. All tested levels belong to the Trimble DiNi 12 series. The third level belongs to the older versions of this series of levels, and it use older version of software than other two levels. The older version of the software also influenced the significant change in corrections for the staff reading near limiting values. Since this level has not been tested in the past, therefore we cannot evaluate the impact of their long-term use.

ACKNOWLEDGEMENT

"This publication was supported by Competence Center for SMART Technologies for Electronics and Informatics Systems and Services, ITMS 26240220072, funded by the Research & Development Operational Programme from the ERDF."

REFERENCES

Woschitz, H.; Brunner, F. K. 2003. Development of a vertical comparator for system calibration of digital levels. *Österreichische Zeitschrift für Vermessung und Geoinformation*, vol. 91, 68–76.

Takalo, M., Rouhiainen, P. (2004). Development of a system calibration comparator for digital levels in Finland. *Nordic Journal of Surveying and Real Estate Research*, 1(2), 121–130.

Vyskočil, Z., Lukeš, Z. (2015). Horizontal comparator for the system calibration of digital levels–realization at the Faculty of civil engineering, CTU Prague and in the laboratory of the Department of survey and mapping Malaysia (JUPEM) in Kuala Lumpur. *Geoinformatics FCE CTU*, 14(2), 55–61.

Ingensand, H. (1999). The evolution of digital levelling techniques-limitations and new solutions. *The importance of heights. FIG*, Gävle, Sweden, 59–68.

Takalo, M., Rouhiainen, P. (2002). On system calibration of digital levels. In: *Proceedings of the XIV General Meeting of the Nordic Geodetic Commission*, 278. 2002.

Gassner, G. L., et al. (2007). Investigation of Leveling Equipment for High Precision Measurements. *Stanford Linear Accelerator Center (SLAC)*.

Wu, Ch. T., Chen, Ch. S., Chang, M. W. (2013). Uncertainties in the Calibration System for Invar Leveling Rods. *Journal of Applied Science and Engineering, 16(1), 69–78*.

Maurer, W., Schnädelbach, K. (1995). Laserinterferometry - Ten Years Experience in Calibrating Invar Levelling Staffs. In: *Proc., First Int. Symp. Appl. Laser Techniques in Geodesy and Mine Surveying, Ljubljana. 1995*.

Advances and Trends in Geodesy, Cartography and Geoinformatics II –
Molčíková, Hurčíková, & Blišťan (eds)
© 2020 Taylor & Francis Group, London, ISBN 978-0-367-34651-5

Measurement and documentation of St. Spirit Church in Liběchov

T. Křemen
Department of Special Geodesy, Faculty of Civil Engineering, Czech Technical University in Prague, Czech Republic

ABSTRACT: We often encounter incompleteness or even a lack of documentation of a given monument in the care of material cultural heritage. In this case, there is no other way than to complete the documentation of the monument or to create a completely new documentation. This was the case of the Church of St. Spirit in Liběchov, which is currently in a poor condition. During the planning of the repairing it was found that there is no drawing documentation for the church. Students of CTU in Prague offered to measure the church on its diploma theses. The church was measured by laser scanning and aerial photogrammetry. During the processing of the measured data, the drawing documentation and 3D CAD model of the whole church were created. The results were handed over to the Mayor of Liběchov as a basis for preparing the project for the planned repair of this cultural monument.

1 INTRODUCTION

While taking care of tangible cultural heritage we often interfere with incompleteness and even with non-existence of the technical documentation of the given monument. The only solution in such a situation is to fill in the technical documentation of the given monument or to create a completely new one. Survey documentation carried out on historical buildings has its own specifics, by which it differs from classical documentation of the real construction typical for new buildings. The measurement should be carried out by a surveyor in cooperation with a building historian or an architect dealing with monuments. It should result from their cooperation for which purpose is the measurement documentation provided, what is expected from it, what detail and truth level the measurement of the given subject should have and what single key elements should be measured (Veselý 2014).

Survey documentation and processing should be carried out using modern geodetic methods as the global navigation satellite systems (GNSS), total stations, laser scanning systems (LSS) and photogrammetry (Štroner et al. 2013, Pukanská 2013, Pavelka 2016a). Using combination of these technologies together with efficient computer and software equipment enables us to obtain a lot of new types of outputs on a higher qualitative level as space models and animations that provide us in combination with classical drawings with very comprehensive view of the given historical objects in a broader context (Křemen 2013, Pavelka 2016b).

That was the case of the Church of Saint Spirit in Liběchov from the 17th century, which is currently in a very poor technical condition. A local volunteer society tries to transfer it to ownership of the Liběchov community and its reconstruction. It was found out during preparation of planning the future reparation of the church that there is no drawing documentation to the church. Students of CTU, Faculty of Civil Engineering, field of Geodesy and Cartography offered to measure the whole church and create its drawing documentation and a 3D model within their theses (Dvořáčková 2018, Klimánková 2018, Valešová 2018). They used combination of LSS and photogrammetry for measurement of the object.

2 ST. SPIRIT CHURCH

The Saint Spirit Church is an early baroque sacral building, which is a dominant of the vineyard hill above the town of Liběchov in the Central Bohemian Region close to Mělník (Figure 1). It is a single nave basilica with a prism tower in the face, which is oriented towards north-west –south-east. The nave has an oblong ground-plan closed inside with polygonal and outside with semi-circular presbytery. There is a crypt under the church (Památkový katalog 2019). The church tower serves as a trigonometric point in whose neighbourhood there are two locking points stabilised by granite boundary marks and a levelling bench mark settled into the exterior presbytery wall.

Figure 1.   St. Spirit church (Dvořáková, 2018).

3 MEASUREMENT

The church was measured in several steps. Surveying net was created before measuring for documentation purposes. Two points of lock of trigonometrical station were used as a base of the surveying net. Other two surveying points were stabilized around the church and next two surveying points were stabilized inside the church. Leica TS06 total station (Figure 2a) was used for measurement. Surveying net was measured by a spatial traverse and it was adjusted using least squares method (Hampacher, 2015). The surveying net was used for georeferencing of the mapping into the S-JTSK coordinate system (Souřadnicový systém Jednotné trigonometrické sítě katastrální; Datum of Uniform Trigonometric Cadastral Network) and the Bpv datum (Výškový systém baltský - po vyrovnání; Baltic Vertical Datum - After Adjustment).

3D terrestrial laser scanning was used as the main method for measurement of the church. 3D scanning system FARO Focus 3D X 330 was used (Figure 2b). Scanner measurement principle is the spatial polar method. FARO has phase electronic distance meter, which measures distances up to 330 m. The field of view is 300° in vertical direction and 360° in horizontal direction. The scanning speed is up to 976000 points per second. The scanner is equipped with a two-axis. The scanner weight is 5 kg. The measurement was started in the crypt of the church then the navy and presbytery were measured. Spiral staircase and three balconies in the first floor were measured at the end of measurement in the interior. Next measurement was carried out outside around the church. There were measured 10 stations in the crypt, 12

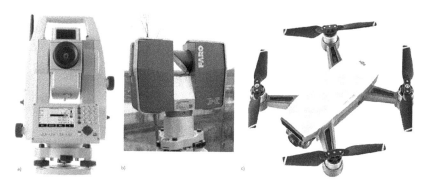

Figure 2.   a) Leica TS06 total station b) FARO Focus 3D X330 c) DJI Spark.

stations in the ground floor, 13 stations in the staircase and balconies and 7 stations outside the church. Five stations were above the surveying points of the surveying net. Following scanning parameters were set: quality of measurement 4 (it is technical parameter), scanning density 12 mm in distance 10 m, time of scanning 2 minutes and 47 seconds for interior and quality of measurement 4, scanning density 6 mm in distance 10 m, time of scanning 8 minutes and 9 seconds for exterior. Measurement was carried out as a traverse with three-tripod system. Ending points of the three-tripod system were signalized by spheres with diameter 200 mm. Other spherical targets with diameter 145 mm were set around the traverse in the ground floor of the navy for better process of registration.

Photogrammetry was used as auxiliary method for measuring upper parts of church. UAV (unmanned aerial vehicle) DJI Spark (Figure 2c) and digital camera Olympus E-300 were used for measurement of the roof and upper parts of the church tower. 29 photogrammetric control points were measured for georeferencing of photogrammetric result. Corners of the church and corners of individual brick were used as control points. Small number of photos was taken for safe model creation because big trees were close to church and UAV navigation signal was disturbed by near transmitter and it was the main reason why so many control points were used.

4 PROCESSING

The first part of processing was registration of the scanned point clouds from laser scanning. Registration was carried out in the Cyclone software. Spherical targets were modelled and numbered in the beginning of the process. Crypt and ground floor were registered into first group, balconies in the first floor were registered into second group and exterior was registered into third group by spherical targets. These three groups of point clouds and point clouds of the spiral staircase were registered to the final coordinate system S-JTSK and Bpv datum in the next step. Point clouds of the spiral staircase were registered by overlaps of the point clouds because it was not possible deploy spherical targets there. The accuracy of the final registration was about 5 mm.

The second part of processing was creation of point cloud from photogrammetry in software Agisoft Photoscan. SFM (Structure from motion) method was used for determination of elements of interior and exterior orientation. Photos were aligned with medium accuracy. Position in the space and dimension was set by photogrammetric control points. The point cloud was built by build dense cloud function with high quality and mild depth filtering. The last step of registration was connection point clouds from laser scanning and photogrammetry. The point clouds from laser scanning, from photogrammetry and final point cloud are in the Figure 3.

Accuracy of the final point cloud was checked by check distances inside and detailed points measured by total station outside. Seventeen check distances were measured and average difference between check distances and distances measured in the final point cloud was 7 mm. Twenty one distances were calculated from detailed points. These distances were spatial. They went through the church from one side to other side. Average difference between distances

Figure 3.   a) point cloud from laser scanning b) point cloud from photogrammetry c) final point cloud (Valešová, 2018).

from total station and distances from point cloud was 14 mm. These average differences contain error from wrong identification of the ending point of the checked distance, because the measured object is in a very poor condition with chipped facades and damage edges. Therefore, it is possible say that accuracy of the final point cloud is high.

2D drawing of the church were created after checking of accuracy. Creation of drawings was started in the Cyclone software. The final point cloud was transformed into local coordinate system of the church for the drawing. Reason for creation of local coordinate system was better manipulation with the model and better definition of cut planes and reference planes for drawing. Axis X goes through the main axis of the church and null of the axis Z was set to the upper surfaces of the entrance stone threshold (Figure 4).

Then the point cloud was cut into small parts and these parts were vectorized. Vectorization was done in reference planes, thereby drawing was in 2D. The vectorization was exported into MicroStation software. 2D drawings were finalized there. Figure 5 shows drawing example.

There were created ground plans of each level of the church, four views to each sides of church in exterior and two longitudinal section and two cross-section. The drawings were carried out mostly according to the standard (ČSN 013420, 2004).

The last step was creation of a 3D CAD (computer-aided design) model of the church. Working on 3D model was started by vectorization in the Cyclone software. Line segments and arcs were drawn directly into the point cloud, where ending points of the linear objects were identical to the point cloud points, or they were drawn into the reference planes. These drawings were exported into MicroStation software. Spatial wired model of the church was made there and after that, surfaces were modelled from line drawing. Figure 6 shows final 3D surfaced model of the church.

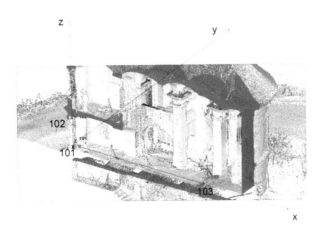

Figure 4.   Sketch of the local coordinate system for 2D drawing (Klimánková, 2018).

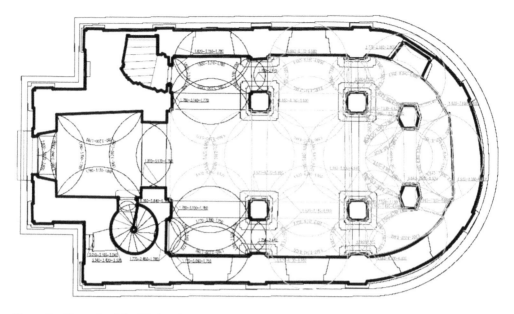

Figure 5. Example of the 2D drawing – ground plan of the nave (Dvořáková, 2018).

Figure 6. View to the 3D CAD model of the whole church (Valešová, 2018).

5 CONSLUSION

Quality surveying documentation of the cultural heritage is an important information for planning repairs and their reconstruction. Nowadays there are many new and progressive measuring technologies. Combination of these technologies produces an accurate and very detailed surveying documentation. The article describes the procedure of measurement and creation of the surveying documentation of the St. Spirit church in Liběchov. Combination of terrestrial laser scanning and aerial photogrammetry was used for measurement. The drawing documentation of the whole church was created during the processing of the measured data. There were the ground plans of the crypt, ground floor and balconies and views of the church interior and exterior. Furthermore, a 3D CAD model of the whole church was created for a better spatial idea of the building and for presentation purposes.

The resulting point cloud, drawing documentation and 3D model were handed over to the mayor of Liběchov as a basis for preparation of the project for the planned repair of this cultural monument, which is in a desolate state. The mayor evaluated the results of our students very positively.

ACKNOWLEDGEMENT

This work was supported by the Grant Agency of the Czech Technical University in Prague, grant No. SGS19/047/OHK1/1T/11 – "Optimization of acquisition and processing of 3D data for purpose of engineering surveying, geodesy in underground spaces and 3D scanning".

REFERENCES

ČSN 013420. 2004. *Výkresy pozemních staveb – Kreslení výkresů stavební části.* Český normalizační institut, Czech Republic.

Dvořáková, K. 2018. *Zaměření a zpracování vybrané výkresové dokumentace kostela sv. Ducha v Liběchově.* Diploma thesis Faculty of Civil Engineering CTU in Prague.

Hampacher, M. & Štroner, M. 2015 *Zpracování a analýza měření v inženýrské geodézii.* Praha: CTU Publishing House, p. 336 ISBN 978-80-01-05843-5.

Klimánková, P. 2018. *Zaměření kostela sv. Ducha v Liběchově a vyhotovení části jeho výkresové dokumentace.* Diploma thesis Faculty of Civil Engineering CTU in Prague.

Křemen, T. & Koska, B. 2013. *2D and 3D Documentation of St. Nicolas Baroque Church for the General Reconstruction Using Laser Scanning and Photogrammetry Technologies Combination.* In: Videometrics, Range Imaging and Applications XII; and Automated Visual Inspection. Bellingham (stát Washington): SPIE, ISSN 0277-786X, ISBN 978-0-8194-9607-2.

Pavelka, K. & Šedina, J. & Housarová, E. & Martan, M. 2016a. *Using multi image photogrammetry for precise documentation of historical building parts.* In: 16th International Multidisciplinary Scientific Geoconference SGEM 2016 Book 2 In-formsatics, Geoinformatics,and Remote Sensing Volume II. Sofia: International Multidiscipli-nary Scientific GeoConference SGEM, pp. 1083–1090, ISSN 1314-2704, ISBN 978-619-7105-59-9.

Pavelka, K. & Hůlková, M. & Matoušková, E. & Havlín, J. 2016b. *Exploration of the Charles bridge in Prague by GPR and TLS technology.* Interdisciplinarity in Theory and Practice, pp. 264–268, ISSN 2344-2409.

Památkový katalog [online] [cit. 2019-06-20] https://pamatkovykatalog.cz/kostel-sv-ducha-14533955.

Pukanská, K. 2013. *3D VISUALISATION OF CULTURAL HERITAGE by using laser scanning and digital photogrammetry.* Ostrava, VŠB - Technická univerzita Ostrava, p .107, ISBN 978-80-248-3214-2.

Štroner, M. & Pospíšil, J. & Koska, B. & Křemen, T. & Urban, R. & Smítka, V. & Třasák, P. 2013 *3D skenovací systémy.* Praha: CTU Publishing House, p. 396, ISBN 978-80-01-05371-3.

Valešová, D. 2018. *Zaměření kostela sv. Ducha v Liběchově a vytvoření prostorového modelu.* Diploma thesis Faculty of Civil Engineering CTU in Prague.

Veselý, J. 2014. *Měřická dokumentace historických staveb pro průzkum v památkové péči.* Praha: Národní památkový ústav. ISBN 978-80-86516-79-0.

Advances and Trends in Geodesy, Cartography and Geoinformatics II –
Molčíková, Hurčíková, & Blišťan (eds)
© 2020 Taylor & Francis Group, London, ISBN 978-0-367-34651-5

Aspects of establishing calibration baselines for electronic distance meters – site selection and configuration of baseline points

M. Papčová
Research Institute of Geodesy and Cartography in Bratislava, Slovakia

J. Papčo
Department of Theoretical Geodesy, Faculty of Civil Engineering, Slovak University of Technology in Bratislava, Slovakia

ABSTRACT: Reliable geodetic measurements are performed by checked, tested and calibrated instruments. Otherwise an incorrect measured value or a measured value with lower accuracy than it is specified by the manufacturer can cause damage a natural person, legal entity and the public. For the purpose of calibration and testing of electronic distance meters, the calibration baselines are being built in outdoor conditions. This paper is dedicated to general aspects that need to be solved when designing the baselines. It contains requirements for the site location of the baselines. Also deals with types of geometric design of the baselines, i. e. with the configuration of the baseline points. The application of these aspects in the past and nowadays is presented by two baselines, which were built thirty years apart – the baseline Hlohovec in Slovakia and currently the most modern baseline in Munich in Germany. The contribution of this paper is the state of the art of building the baselines.

1 INTRODUCTION

Geodetic activities are performed reliably only by correctly functioning instruments, by instruments that meet the accuracy specifications given by manufacturers and by instruments with known instrumental and non-instrumental systematic errors. For the reason it is necessary to perform checking, testing and calibration of instruments at regular intervals (Heister 1999).

For electronic distance meters (EDM), the following systematic errors of measured distances can occur (Rüeger 1996): zero point error, scale error, short periodic errors (cyclic error), nonlinear distance-dependent errors (as phase inhomogeneity). There are already known procedures and devices to determine these errors (Joeckel et al. 2008, Rüeger 1996). This article focuses on the first two mentioned systematic errors and on one of the devices for calibration and testing electronic distance meters - the EDM baseline in the outdoor environment.

Two papers in this proceeding cover the aspects to be addressed and taken into account when establishing EDM baselines. This paper focuses on the location and configuration of baseline points. Each aspect is described first by general principles that are further specifically applied to two baselines. The description relates to the baseline Hlohovec in the Slovak Republic (year of construction 1978-1979) and the most modern baseline Munich (year of construction 2009) that were built with thirty years delay.

These papers review a comprehensive state of the art of the establishing EDM baselines. The structure of this paper consists of three further chapters. The second chapter lists the main goals of the baseline. The third chapter describes the requirements that are favorable to the baseline placement. The following chapter is devoted to a type of baseline's geometric designs and choice of design parameters for the calculation of baseline's section.

2 MAIN PURPOSES OF THE EDM BASELINE

The EDM baseline consists of linearly arranged, permanently stabilized pillars and serves to:

– determinate the correction of the constant part of the zero point error and the scale correction which includes a frequency error and a distance dependent part of the zero point error if the lengths between the baseline points are known and traceable to the SI meter definition (Rüeger 1996, Joeckel et al. 2008),
– determinate the constant part of the zero point error if the lengths between the baseline points are not known (Rüeger 1996, Joeckel et al. 2008),
– determinate the experimental standard deviation of a single distance measurement, determinate the zero point correction and its standard deviation in according to ISO 17123-4, baseline's section lengths are unknown.

Zero point error EDM is defined as the difference between geometric (should be) and measured electronic distance (is) (Deumlich et al. 2002). Zero point error is generally a distance-dependent parameter. The total zero point error can be divided into a constant part at D = 0 m and a distance dependent part (liner, polynomial) (Joeckel et al. 2008). According to (Schwarz 2012), there is a distance-dependent part only in rare cases. The distance dependence is caused by the phase inhomogeneity of the transmitting diode (Witte et al. 2006, Schwarz 2012), dependence on the EDM temperature (Rüeger 1996), voltage dependence or signal intensity (Rüeger 1996, Schwarz 2012). The zero point error is given to the combination of distance meter and reflector (Witte et al. 2006, Deumlich et al. 2002), and their influence cannot be separated (Schwarz 2012).

The scale error is mainly caused by the variation in the modulation frequency of phase EDM and counting frequency in pulse EDM (Joeckel et al. 2008) due to aging of the quartz and temperature dependence of oscillator (Schwarz 2012). Scale error represents the distance deviation directly proportional to the measured distance.

3 SITE SELECTION

The EDM baseline location must be selected carefully (Surveying JRP-Consortium 2016) to meet its main goal - the determination of the parameters with low measurement uncertainty. The requirements for the locality suitable for a baseline are summarized in the literature from 1996 to 2016. They do not change during this period; they are only supplemented with regard to the use of EDM baseline for GNSS measurements.

The baseline area is characterized by features that are divided into five main groups:

1. Terrain
   – has a suitable length equal to total baseline length (Ellis et al. 2013),
   – is a shady place for EDM (Surveying JRP-Consortium 2016) or has a free horizon for GNSS (Surveying JRP-Consortium 2016),
   – has a planar (slightly concave - light sag in the middle) ground surface (horizontal or equally oblique) (Rüeger 1996), which ensures the good visibility between pillars (Surveying JRP-Consortium 2016),
   – is covered with even growth (Rüeger 1996) which has apositive impact on meteorological conditions, e.g. low grass, not asphalt (Joeckel etal. 2008),
   – has anorth-south orientation allowing measurements in and against the direction of the Sun (Rüeger 1996),
   – clear of overhead obstructions and surrounding vegetation (Ellis et al. 2013),
2. Geology
   – is geologically stable with a homogeneous subsoil to guarantee the long-term stability of pillars (Surveying JRP-Consortium 2016), (Joeckel et al. 2008), e.g. bedrock (Rüeger 1996),

3. Meteorology
   - is characterized by the same meteorological conditions (wind, sunshine) (Joeckel et al. 2008), e.g. by the same sunlight (no combination of shielded and unshielded parts) (Rüeger 1996),
   - located on the embankment representing nearly ideal weather conditions and has a clear view (Heister 2012),
   - with low wind, resulting in low air turbulence (Surveying JRP-Consortium 2016),
4. Transport
   - good transport accessibility (Rüeger 1996), (Joeckel et al. 2008),
   - with a service road located alongside the baseline, thereby achieving faster transport of the reflector and the instrument between the baseline's points (Rüeger 1996),
   - with low traffic (Joeckel et al. 2008),
5. Other
   - has limited or non-public access due to the risk of vandalism and interruption of the measuring beam (Rüeger 1996),
   - without the effects of human activity in the neighborhood, e.g. on building site or transport (Surveying JRP-Consortium 2016),
   - without planned changes in the future - construction, expansion of the road, drainage, etc. (Ellis et al. 2013),
   - whose landowner supports the construction of a baseline (Rüeger 1996).

(Ellis et al. 2013) considers the site selection as a time-consuming process and also pointing out the obstacles for which the chosen site can be rejected.

The baseline Hlohovec is situated on a flat area with a small slope in the direction of the river Váh with originally homogeneous vegetation (Priam et al. 1992). The subsoil consists of sand-gravel sediments (Abelovič et al. 1980). Baseline points should be stabilized below the groundwater level (Abelovič et al. 1980), whilst it cannot drop below the baseline points (Priam et al. 1992). However, control measurements indicate that below the groundwater level there is a lower part of only two pillars 1 and 5, which may result in significant height and position changes of points 2 to 4. It is also possible to observe conspicuously the decreasing of points 3 and 4 after the construction of the sewage treatment plant and the excavation of the waste channel (Mitáš et al. 2002). The baseline points are reachable through the field path along the left bank of the river Váh (Priam et al. 1992).

The baseline in Munich is located in the flat terrain of the Universität der Bundeswehr with a free horizon (Heister 2012). The terrain is geologically stable within the Munich gravel plain (Heunecke 2012). To achieve almost ideal meteorological conditions, the terrain was picked up by creating a 2 m high embankment (Heister 2012). The baseline is well accessible by taxiway of old airport (Heister 2012). The advantages of the selected fenced area are the unavailability for the public and its protection.

The selection of locations for both baselines is affected by the same requirements. In Hlohovec there was impossible to arrange stable subsoil and to limit further activity in the site. The baseline Munich has a new element – embankment for better atmospheric conditions.

4 CONFIGURATION OF BASELINE POINTS

The baseline consists of a group of points that are arranged in a design in according to certain instructions. The main features of several types of baseline designs are described in subchapter 4.1. Subchapter 4.2 is addressed to the choice of design parameters for the calculation of the section lengths of the most used design.

4.1 Types of designs

(Jokela 2014) and (Rüeger 1996) distinguish four types of baseline designs:

1. design of older baselines,
2. Heerbrugg design,

3. Aarau design,
4. Hobart design.

The idea of design of older baselines has been adapted to the baseline calibration measurement, using 24 m invar wires. Therefore, section lengths are formed by multiples of 24 m (Jokela, 2014).

Heerbrugg design, also called Schwendener design (Rüeger 1996)

– includes a uniform length distribution along the total baseline length as well as along the unit length of the distance meter (U),
– lengths or fractions of U are not repeated,
– allows detection of all distance-dependent errors, including short periodic errors,
– measurement is performed in all combinations.

Short periodic errors must be checked using the residual graph against the corresponding unit length U. If they are present, the adjustment must be repeated with the corrected data on short periodic error, otherwise it may be flawed additive constant up to 75% of the amplitude of short periodic errors (Rüeger 1996). It is more suitable to include short periodic errors as unknown parameters in the adjustment (Rüeger 1996).

Aarau design (Rüeger 1996)

– includes uniformly distributed lengths over the total baseline length,
– does not detect short periodic errors, although short periodic errors are determined separately on a special cyclic error test line (Rüeger 1996, Joeckel et al. 2008),
– has baseline section of the exact multiples of U,
– is measured in all combinations.

If short periodic errors are not determined beforehand, they affect each combination measured on the baseline with the equal amount. As a result, short periodic errors are only transmitted to the additive constant. The determined scale or unknown section lengths of the baseline will not be affected (Rüeger, 1996).

Hobart design (Rüeger 1996)

– has pillars placed at uniform intervals along the baseline range,
– is not based on distance measurements in all combinations,
– must have known section lengths between pillars,
– uses only one endpoint of the baseline as a measuring station as well as an auxiliary station shifted at U/2,
– is measured in minimum time because the distance meter is moving only a few meters,
– is characterized by poor redundancy, which can be improved by a larger number of baseline stations and/or other auxiliary station on ¼ and ¾ U from the first station.

Both parameters are not affected by first-order short periodic errors and at least third-order short periodic errors, but are affected by second order short periodic errors (Rüeger 1996).
Newer designs (2.-4.) are based on the way of handling the cyclic error as its determination or its elimination. The advantage of Heerbrugg design compare to Aarau and Hobart design is the determination of cyclic error directly on the outdoor baseline. It should be noted that today's accurate instruments generally have a cyclic error amplitude of up to approx. 0.1 mm (Heunecke 2012). This means that even with the Heerbrugg type, a cyclic error is determined at the longitudinal baseline (Heunecke 2012) in indoor conditions.
Section lengths of individual designs are calculated according to the design equations given in (Rüeger 1996).

4.2 *Selection of design parameters*

The latest baselines (published Heister 2012, Jokela et al. 2010, Ellis et al. 2013) such as Munich, Innsbruck, Eglinton and also the test line setup for EDM in ISO 17123-4 and DIN

18723-6 use Heerbrugg design. Four design parameters, which are discussed in the following subchapters, are involved in the equations of this design type.

4.2.1 *Unit length*
Although only one U enters into the design equations, it is necessary to check the uniform section length distribution along the different U of today's instruments (Victoria State Government 2019) as in (Heister 2012, Ellis et al. 2013) when choosing the resulting design version.

4.2.2 *Shortest distance*
The constant and the linear term are usually estimated during the calibration. (Joeckel et al. 2008) points to strong nonlinear zero point error in the close range. Significant distance differences occur in the close range up to max. 10 m (Juretzko 2019). When testing a constant part of the zero point error, (Deumlich et al. 2002) it is recommended that the section lengths should be at least 30 to 50 m long because the zero point error is distance-dependent especially at short distances up to 50 m. As consequence, the shortest baseline distance should be chosen the distance without the unwanted effect of the zero point error. Simultaneously the shortest distance should be a multiple of the unit length according to (Rüeger 1996).

4.2.3 *Total baseline length*
The total length of baseline has to be equal to

1. the smallest distance range of instruments (Rüeger 1996),
2. the largest distance range of instruments, while it should be built more points to determine the parameters of the instrument with a smaller distance range (Rüeger 1996),
3. the typical distance range of practical geodetic activity (Rüeger 1996), (Joeckel et al. 2008); (Surveying JRP-Consortium 2016) recommends a longer baseline is suitable for determining the scale factor of low-uncertainty.

It is very difficult to find a suitable location for the whole distance range of the instruments in according to the first and second options. In case of long baseline it is hard to achieve homogeneous meteorological conditions because of increasing uncertainty of determining meteorological parameters. The third option is applied to the current baselines.

Literature recommends a total baseline length of 500 to 2200 m (Deumlich et al. 2002 - 800 to 1000 m, Witte et al. 2006 - 640 m, Kahmen 2006 – less than 1000 m, Surveying JRP-Consortium 2016 - 500 - 1000 m, ISO 17123-4 - 300-600 m, DIN 18723-6 500-600 m). The established baseline in Europe has the total length from 204 to 2202 m, in the USA up to 1500 m (NGS 2019), in Australia from 429 to 888 m (Ellis et al. 2013, planned extension).

4.2.4 *Number of baseline stations*
The number of pillars is chosen as a compromise between the incurred costs and the required uncertainty of the determined distances. Literature mentions 6-8 pillars (Rüeger 1996 - 6-7, Deumlich et al. 2002 - 6, Surveying JRP-Consortium 2016 - 6-8, ISO 17123-4 - 7, DIN 18723-6 - 7). The existing baselines consist of 6-12 pillars. In many cases, the baselines have more pillars, but they commonly use seven pillars like Hannover, Munich, Koštice. Additional pillars are used in case of calibration required in larger distance range.

4.2.5 *Hlohovec, Munich*
The baseline in Hlohovec is formed by mutual distances of the pillars, which are equal to the whole multiple of the measuring modulation frequency of the EDM 10 m and the whole multiple of the invar wire length of 24 m. It is a combination of older design and Aarau design. The minimum length is 120 m and the maximum length of 2550 m, for common testing usually 960 m length is used. It is consisting of seven pillars, but only five pillars were used. The point 6 could be used to test distance meters with the distance standard deviation lower than 5 mm (Abelovič 1985). The pillars are located at the length (Bučko 1989): 1-2 ca. 120 m, 1-3 ca. 240 m, 1-4 480 m, 1-5 960 m, 1-6 1780 m, 1-7 2550 m.

The Munich baseline is of Schwendener type. The design is calculated for U = 3 m and takes into account another unit length of 5 m, 10 m, 2 m and 0.3 m. This design type meets the first seven pillars in the range of up to 590 m, the 8th pillar is shifted by 510 m in a straight line to the 7th pillar. The 8th pillar is used to determine the scale correction of longer distances and for the GNSS measurement. This design type has been already proved by the previous Munich baseline. The pillars are at the distances from the first station 18.8 m, 101.2 m, 247.4 m, 425.4 m, 539.7 m, 590.3 m, 1100.0 m.

Although the baselines have different design, they have met their purpose. The cyclic error is detectable on the newer baseline only theoretically. The newer baseline has more redundant measurements. The validity of the parameters is given by the total baseline length (Woschitz et al. 2018) - Hlohovec (120-960 m), Munich (usually 18.8-590 m, possibly up to 1100 m).

5 CONCLUSIONS

The EDM calibration baseline is used for EDM calibration outside the close range in outdoor conditions. The basis for establishing the baseline is the selection of a suitable territory. Terrain, geology, meteorology, transport, and other phenomena are taken into account in process of selecting a territory. In spite of the requirements, the selected territory represents a compromise between ideal requirements and real conditions. The configuration of baseline points has four well-known designs: design of older baselines, Heerbrugg, Aarau and Hobart design. The input parameters should be considered for calculating the section lengths. The calibration results will refer to the measured distance range. The problems mentioned above, the new element and realization of specific baselines can warn or motivate us when designing a new baseline. Other aspects taking into account when establishing a baseline are listed in the second paper of this proceeding.

REFERENCES

Abelovič, J. & Jurda, K. 1980. Testovacia základnica pre elektronické diaľkomery. *Geodetický a kartografický obzor* 26/68(10): 243–247.
Abelovič, J. 1985. Súčasný stav a perspektívy využívania testovacej základnice v teréne. In Abelovič, J. (ed.) *Testovanie diaľkomerov na geodetickej základnici v teréne.*
Bučko, E. 1989. Určenie základných charakteristík presnosti elektronických diaľkomerov. *Geodetický a kartografický obzor* 35/77 (6): 143–146.
Deumlich, F. & Staiger, R. 2002. *Instrumentenkunde der Vermessungstechnik.* Heidelberg: Wichmann.
DIN 18723-6, 1990. Field procedure for precision testing of surveying instruments; electro-optical distance measuring instruments for short ranges.
Ellis, D. & Janssen, V. & Lock, R. 2013. Improving Survey Infrastructure in NSW: Construction of the Eglinton EDM Baseline. In *Proceedings of the 18th Association of Public Authority Surveyors Conference,* Canberra, Australia.
Heister, H. 1999. Checking, Testing and Calibrating of Geodetic Instruments – Some remarks with respect to recent developments in this field. In Lilje, M. (ed.) *Geodesy and Surveying in the Future – The Importance of Heights,* Gävle, Sweden.
Heister, H. 2012. The new Calibration Baseline of the UniBw Munich. *Allgemeine Vermessungsnachrichten* 119 (10): 336–343.
Heunecke, O. 2012. Die neue Neubiberger Pfeilerstrecke. *Zeitschrift für Geodäsie, Geoinformation und Landmanagment.* 140(6): 357–364.
STN ISO 17123-4. 2013. Optics and optical instruments – Field procedures for testing geodetic and surveying instruments – Part 4: Electro-optical distance meters (EDM instruments to reflectors).
Joeckel, R. & Stober, M. & Huep, W. 2008. *Elektronische Entfernungs- und Richtungsmessung und ihre Integration in aktuelle Positionierungsverfahren.* Heidelberg: Wichmann.
Jokela, J. & Häkli, P. & Kugler, R. & Skorpil, H. & Matus, M. & Poutanen, M. 2010. Calibration of the BEV Geodetic Baseline. In *FIG Congress 2010, Facing the Challenges – Building the Capacity*, Sydney, Australia.
Jokela, J. 2014. Length in Geodesy – on metrological traceability of a geospatial measurand. PhD thesis.

Juretzko, M. 2019. Geodätisches Institut, Karlsruher Institut für Technologie. Personal communication.

Kahmen, H. 2006. *Angwandte Geodäsie: Vermessungskunde*. Berlin: Walter de Gruyter.

Mitáš, J. & Mičuda, J. & Bučko, E. 2002. Výškové zmeny na geodetickej porovnávacej základnici Hlohovec. *Geodetický a kartografický obzor* 48/90(10): 185–188.

NGS (National Geodetic Survey). 2019. https://www.ngs.noaa.gov/CBLINES/calibration.shtml.

Priam, Š. & Blahunka, J. & Hajdúch, Ľ. 1992. Výskumná úloha č. M 21/92/EO Zabezpečenie podkladov na vyhlásenie geodetickej dĺžkovej zákaldnice Hlohovec za súčasť štátneho skupinového etalónu veľkých dĺžok.

Rüeger, J.M. 1996. *Electronic distance measurement. An Introduction*. Berlin Heidelberg: Springer.

Schwarz, W. 2012. Influence Qualities for Electro-Optical Distance Measurements and their Acquisition. *Allgemeine Vermessungsnachrichten* 119 (10): 323–335.

Surveying JRP-Consortium. 2016. Good practice guide for the calibration of electro-optic distance meters on baselines.

Victoria State Government. 2019. https://www.propertyandlandtitles.vic.gov.au/surveying/survey-equipment-calibration

Witte, B. & Schmidt, H. 2006. *Vermessungskunde und Grundlagen der Statistik für das Bauwesen*. Heidelberg: Wichmann.

Woschitz, H. & Heister, H. 2018. Überprüfung und Kalibrierung der Messmittel in der Geodäsie. In Schwarz W. (ed.) 2018. *Ingenieurgeodäsie*. Berlin: Springer Spektrum.

Advances and Trends in Geodesy, Cartography and Geoinformatics II –
Molčíková, Hurčíková, & Blišťan (eds)
© 2020 Taylor & Francis Group, London, ISBN 978-0-367-34651-5

Aspects of establishing calibration baselines for electronic distance meters – position and height design, point monumentation and metrology measurement

M. Papčová
Research Institute of Geodesy and Cartography in Bratislava, Slovakia

J. Papčo
Department of Theoretical Geodesy, Faculty of Civil Engineering, Slovak University of Technology in Bratislava, Slovakia

ABSTRACT: Reliable geodetic measurements are performed by checked, tested and calibrated instruments. Otherwise an incorrect measured value or a measured value with lower accuracy than it is specified by the manufacturer can cause damage a natural person, legal entity and the public. For the purpose of calibration and testing of electronic distance meters, the calibration baselines are being built in outdoor conditions. This paper is dedicated to general aspects that need to be solved when designing the baselines - to positional and height design of the baseline, to monumentation of baseline points and to metrology measurement at the baseline. The application of these aspects in the past and nowadays is presented by two baselines, which were built thirty years apart – the baseline Hlohovec in Slovakia and currently the most modern baseline in Munich in Germany. The contribution of this paper is the state of the art of establishing the baselines.

1 INTRODUCTION

This article is a continuation of the article Aspects of establishing calibration baselines for electronic distance meters – site selection and configuration of baseline points which is also in this proceeding. The next aspects are described in same style as in the first article – first generally and then the application for two baselines.

This article is structured into three main chapters: the second chapter deals with positional and height design, the third chapter describes the monumentation of baseline points and the following chapter presents the measurement methods of meteorological parameters at the baseline.

2 POSITION AND HEIGHT DESIGN

The baseline points are arranged into a spatial (3D) line regarding to the equipotential surface or to the plane of the local coordinate system.

The baseline points have to be stacked out in a straight line with sufficient uncertainty to ensure that their position and height deviations do not affect the lengths in the line.

The geometric reductions of measured distances should be considered in case of designing, especially for baseline point heights.

The most convenient way of defining the pillar heights is without any respectively/or with minimal geometry reductions in distance during the distance processing, e. g. during the EDM calibration (simplicity, lower uncertainty) (Eschelbach et al. 2015). The reduction from tilt and height at the baseline is significant regarding the known geometric reductions. Two ways of this distance geometry reduction are applied:

1. to the equipotential surface, which passes through the mean height of the pillars (or mean height of the measuring beam) as in (Figure 1, Distance 2, but at a different height level) (Eschlebach et al. 2015),
2. to the horizontal xy plane of the local coordinate system that passes through the middle height of the pillar plates (Heunecke 2012); the height of the pillars is determined by the geometric leveling, taking into account the curvature of the Earth when converting the leveled height difference into a rectangular coordinate system (Figure 1, Distance 1).

The Hlohovec baseline points lie in a straight line (both directionally and vertically) in one horizontal plane with sufficient accuracy (Priam et al. 1992). Existing points have offset from the horizontal direction max. 0.058 m and from the vertical direction max. 0.26 m. Due to the section lengths of this baseline, deviations from the spatial line were negligible. Distances refer to the baseline axis.

All pillar heads of the Munich baseline are placed high accurate in one horizontal and vertical plane, uncertainty of the horizontal alignment is 3 mm and vertical alignment of 3 mm (Heister 2012). Pillar heights were designed with minimal distance geometric reduction from tilt and height (according to the second method) (Heunecke 2012). With the equal height of the instrument and reflector height (0.335 m), the reduction from tilt and height reaches max. 0.06 mm (Heunecke 2012).

Both baselines are characterized with the same positional and height designs. Different requirements to the pillar alignment uncertainty are resulted from different section lengths.

3   MONUMENTATION OF BASELINE POINTS

The baseline points were marked by ground markers (Rüeger 1996) in the past, but now exclusively by the pillars (JRP-Consortium 2016). The pillar benefits are (Rüeger 1996):

- centering uncertainty of 0.1 mm or better,
- constant height of the instrument and reflector,
- stability during measurement (effects of operator and sun),
- no risk of shifting when replacing the reflector and heavy instrument.

Surveying JRP-Consortium (2016) recommends the construction of pillars according to (Eling et al. 2014), with the following approach of monumentation: The point monumentation

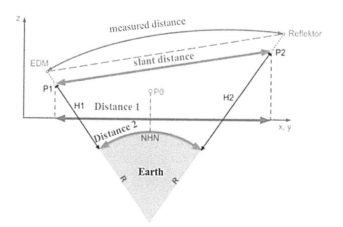

Figure 1.   Geometry reduced distances (Schwarz 2012).

consists of the underground part and above ground part. The first one is a concrete foundation, its depth depends on the geological bedrock, but at least below the frost-resistant foundation (about 80 cm below ground level). The above-ground part - the inner and outer tubes with a diameter of 30-40 cm is mounted in the foundation and reaches a height of approx. 1.3 m above ground. The inner tube is filled with concrete, the gap between the inner and outer tubes is filled with an insulating layer. Thermal insulation reduces the impact of temperature differences on the pillar. The pillar is closed by a center plate, which is firmly attached to the concrete in the inner tube and is horizontally mounted. The pillar plate is protected against weather influence and vandalism by a cover. On the pillar there is a height benchmark to determine the height of the EDM. Pillars should be built early before they are set in order to consolidate the foundation.

Point monumetation in Hlohovec according to (Priam et al. 1992) (Figure 2): The baseline point is made of casing - steel tube filled with concrete. Its underground part extends to a depth of 2-4 m, the height of the above-ground part is 1.3 m. A smooth stainless steel plate with a central screw (metric thread) for forced centering is attached to the upper end of the casing. An additional adapter allows forced centering of Wild (Witword thread) or Kern type too. The entire head is protected by a lockable steel cover. The height benchmark is welded to the casing.

Point monumetation in Munich according to (Heister 2012) (Figure 3): The underground part is made of a bored pilot depth of 5-6 m with a diameter of 60 cm. Each pilot is enclosed with a reinforced concrete foundation on which a measuring pillar is set up in a separate building section. All parts are fixed with armoring. The above-ground part is made according to (Eling et al. 2014) by combination of concrete pillar and plastic jacket (PVC-tube). The gap between them is filled with thermal insulation (glass pearls). The pillar is closed with a special universal center plate of V4A- steel with a diameter of 42 cm.

The pillar in Munich is protected against the surrounding thermal differences by thermal insulation, unlike the Hlohovec baseline. Both types of pillars are suitable for devices with different threads. Due to the development of instruments, at the baseline in Munich is also possible to fit a reflector for a laser tracker. The monumentation of the baseline point in Munich consists of several parts in a vertical direction, which allows differentiation of the uncertainty of the individual parts (Heister 2012). The greatest demands are required on the uncertainty of the centering plate.

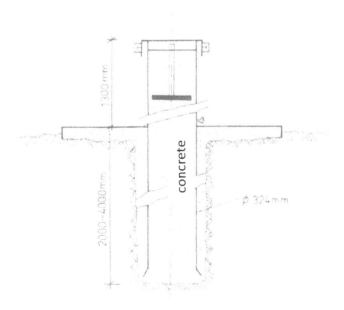

Figure 2.   Cross section of the baseline point in Hlohovec (Priam et al. 1992).

59

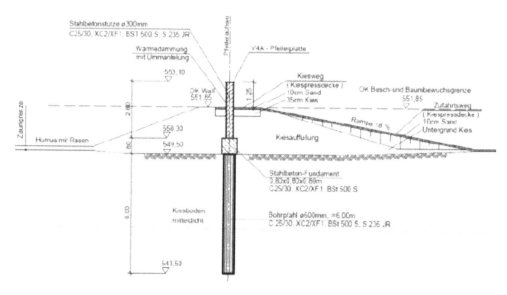

Figure 3. Cross section of the baseline point in Munich (Heister 2012).

4 MEASUREMENT OF METEOROLOGICAL PARAMETERS

The atmospheric conditions along the measuring beam have the greatest impact to the long distances measurements at the baseline (Herrmann et al. 2012, Wasmeier 2012, Nindl et al. 2013). Therefore, it is necessary to consider the way of measuring meteorological parameters to establish the requisite infrastructure for measuring atmospheric parameters when constructing a baseline. It is essential for the latest method of measurement that using a network of wired sensors (wireless solution is problematic). Additional completion may break the baseline point's stability (Pollinger et al. 2012). Surveying JRP-Consortium (2016) recommends establishing a sensor network.

In the next subchapter, we clarify the importance of measuring meteorological parameters. The second subchapter summarizes the ways of measuring the air temperature at the baseline, the third subchapter mentions the ways of measuring pressure and relative humidity and the last subchapter describes its use at both observed baselines.

4.1 Significance

The measured distances at the outdoor baseline are strongly influenced by the propagation velocity of electromagnetic waves in the atmosphere, which depends on the air density or the refractive index of the air (Schwarz 2012, Herrmann et al. 2012, JRP-Consortium 2016 Surveying). This effect on distance measurement is included in the first velocity correction (Schwarz 2012). For its use, it is necessary to measure representative atmospheric parameters along the measurement signal path - temperature, pressure, relative humidity.

It should be noted that an air temperature deviation of 0.1 K, a pressure deviation of 0.34 hPa, a water vapor pressure deviation of 1.67 hPa and a relative humidity deviation of 10% causes distance deviation of 0.1 ppm (Schwarz 2012).

The temperature measurement has the greatest influence on distance measurement. To achieve the distance uncertainty of 0.1 ppm is required to measure air temperature with uncertainty of 0.1 K (Pollinger et al. 2012).

The scale factor of known baseline lengths must be determined with an uncertainty of few tenths ppm (Heister 2012), and the temperature uncertainty of 0.1 K has already caused the distance uncertainty of 0.1 ppm (Pollinger et al. 2012). So, the temperature measurement along the beam represents a very challenging task.

The determination of atmospheric conditions, respectively the first speed correction will be ideal with an integral measurement of meteorology along the measuring beam. But nowadays, it is only possible to place sensors at discrete locations (Nindl et al. 2013). The distance measurements mainly depend on the representativeness and accuracy of meteorological parameters (Eschelbach et al. 2015). The imperfect knowledge of the index of refraction limits the uncertainty of a distance measurement over longer distances (Pollinger et al. 2012).

For example in Braunschweig (Germany), sixty temperature sensors are uniformly distributed at a 600 m long baseline and the temperature changes max. 4.8 K and in average of 1.7 K at one time point (from one year, observe one day per month) (Papčová 2019).

A transversal temperature gradient at the baseline between the centerline and outside the measurement axis (0.7 m) reaches up to 1 K on the sunny day and the difference is below 0.1 K during the cloudy day (Pollinger et al. 2012). At the baseline Munich up to 590 m, the temperature varies on average of 1-2 K (one day) (Heister 2012).

4.2   *Temperature measurement*

The following temperature measurement methods were used for calibration measurements of the baselines:

1. Temperature measurement at both ends of the measured distance (Dvořáček 2017, Wasmeier 2012, Nindl et al. 2013, Jokela et al. 2010, Neuner et al. 2012). This is the minimum requirement for the measurement (at the pillar with instrument and with reflector) and subsequent averaging of measured data (interpolation) (Surveying JRP-Consortium 2016).
2. Temperature measurement at both ends of the measured distance and densify the sensors in some part of the baseline (Neuner et al. 2012). An automatic temperature measuring system was used consisting of 9 sensors located in the middle of the baseline. The mean refractive index is calculated as the weight average depending on the distance between the measuring stations.
3. Temperature measurement on multiple sections of the measured distance (using laser tracker) (Eschelbach et al. 2015). Each distance was divided into four equal sections and a meteorological station was placed in the middle of each section. Or Herrmann et al. (2012) used five sensors for one measured distance by a laser tracker.
4. Network of uniformly distributed sensors along a baseline
   a) a network of 20 sensors along the distance of 1100 m (Herrmann et al. 2012),
   b) a network of permanently marked by 60 sensors over the distance of 600 m (Pollinger et al. 2012).

Uncertainties in the order of magnitude of 0.1 K can be achieved over distances of several hundred meters only under optimal conditions (cloudy sky, light wind) (Meiner-Hagen et al., 2012, Schwarz 2012). In changing lighting conditions, especially in direct sunlight exposure, such uncertainty can not be achieved with conventional sensors (start and end-point of the distance).

The uncertainty of the temperature measurement reaches up to 0.8 K on sunny days, below 0.3 K on sunny days in shadow, below 0.15 K on cloudy days using the sensor network (60 sensors over 600 m) at the baseline in Braunschweig (Pollinger et al. 2012).

4.3   *Pressure and relative humidity measurement*

Measurement of air pressure and relative humidity is unproblematic. The following approaches were applied at baselines:

1. Measurement at the station point (Dvořáček 2017).
2. Measurement at the both end points of the measured distance (Neuner et al. 2012, Wasmeier 2012) – dry, wet temperature and pressure were synchronously measured at both end points of the measured distance.

3. Measurement of relative humidity are carried out at the both end points, pressure measurement only at the station point (Nindl et al. 2013), (Jokela et al. 2010).
4. Other solution, e. g. (Meiners-Hagen et al. 2012) six sensors for measurement of relative humidity and two pressure sensors are placed along the baseline.

4.4 *Hlohovec, Munich*

Meteorological parameters were measured at the start and end point of the measured distance for the calibration of EDM at both baselines.

A sensor network with wireless communication has been installed in Munich, but it has not provided a stable solution, so they recommend sensor cable connection (Liebl 2018).

5 CONCLUSIONS

The baseline points ideally are placed in one spatial (3D) line with sufficient accuracy. The heights of the baseline points are specifically designed with constraint - the geometric reduction from the tilt and height is minimum in order to avoid additional reductions in EDM calibration. The current monumentation of the baseline point reduces the impact of sunlight on the pillar and it is equipped with a universal center plate for electronic distance meters (Wild, Kern) and laser trackers. In case of measuring meteorological parameters by a sensor network, it is necessary to establish its construction simultaneously with the building of the baseline.

This summarizing review of establishing the EDM baseline will serve as a basis for the design of a planned new baseline in the Slovak Republic.

REFERENCES

Dvořáček, F. 2017. Kalibrace terénních délkových základen laserovým trackerem Leica AT401. ČVÚT, Praha, PhD thesis.
Eling, B. & Kemper-Böninghausen, R. 2014. Merkblatt 8-2014. Vermessungspfeiler. https://www.dvw.de/sites/default/files/merkblatt/daten/2015/08_DVW-Merkblatt_Vermessungspfeiler.pdf
Eschelbach, C., Heckmann, B. & Lösler, M. 2015. Determination of High Precision Reference Distances of the Calibration Baseline Neu-Isenburg Using a Mobile Laser Tracker. *Allgemeine Vermessungsnachrichten* 122 (3): 95–101.
Heister, H. 2012. The new Calibration Baseline of the UniBw Munich. *Allgemeine Vermessungsnachrichten* 119 (10): 336–343.
Hermann, Ch., Liebl, W. & Neumann, I. 2012. Laser Tracker Measurements of the KIT Karlsruhe and the UniBw Munich for Determining the Nominal Distances of the Base Line of the UniBw Munich. *Allgemeine Vermessungsnachrichten* 119 (8/9): 309–313.
Heunecke, O. 2012. Die neue Neubiberger Pfeilerstrecke. *Zeitschrift für Geodäsie, Geoinformation und Landmanagment*. 140(6): 357–364.
Jokela, J., Häkli, P., Kugler, R., Skorpil, H., Matus, M. & Poutanen, M. 2010. Calibration of the BEV Geodetic Baseline. In *FIG Congress 2010, Facing the Challenges – Building the Capacity*, Sydney, Australia.
Liebl, W. 2018. Personal communication.
Meiners-Hagen, K. & Pollinger, F. 2012. Traceable Measurements of Long Distances in the PTB. *Allgemeine Vermessungsnachrichten* 119 (8/9): 283–290.
Neuner, H. & Paffenholz, J-A. 2012. Determination of the Baseline from the UniBw Munich with Mekometer ME 5000 - Contribution of the Geodetic Institute from the LU Hannover. *Allgemeine Vermessungsnachrichten* 119 (10): 344–346.
Nindl, D., Hardegen, W. & Wippel, E. 2013. Report about the Cooperative Test on the Calibration Baseline of the UniBw München – Measurement Campaign Leica Geosystems. *Allgemeine Vermessungsnachrichten* 120 (1): 9–18.
Papčová, M. 2019. Analysis of temperature data at the baseline in Braunschweig (not published).
Pollinger, F. Meyer, T., Beyer, J., Doloca, N. R., Schellin, W., Niemeier, W., Jokela, J., Häkli, P., Abou-Zeid, A. & Meiners-Hagen, K. 2012. The upgraded PTB 600 m baseline: a high-accuracy reference for

the calibration and the development of long distance measurement devices. *Measurement science and technology 23.*

Priam, Š., Blahunka, J. & Hajdúch, Ľ. 1992. Výskumná úloha č. M 21/92/EO Zabezpečenie podkladov na vyhlásenie geodetickej dĺžkovej zákaldnice Hlohovec za súčasť štátneho skupinového etalónu veľkých dĺžok.

Rüeger, J.M. 1996. *Electronic distance measurement. An Introduction.* Berlin Heidelberg: Springer.

Schwarz, W. 2012. Influence Qualities for Electro-Optical Distance Measurements and their Acquisition. *Allgemeine Vermessungsnachrichten* 119 (10): 323–335.

Surveying JRP-Consortium. 2016. Good practice guide for the calibration of electro-optic distance meters on baselines.

Wasmeier, P. 2012. Determination of Reference Distances of the Calibration Baseline at the UFAF Munich – Mekometer ME5000 Measurements of the Chair of Geodesy of the Technische Universität München. *Allgemeine Vermessungsnachrichten* 119 (8/9): 305–308.

Advances and Trends in Geodesy, Cartography and Geoinformatics II –
Molčíková, Hurčíková, & Blišťan (eds)
© 2020 Taylor & Francis Group, London, ISBN 978-0-367-34651-5

The determination of deforestation rate in eastern Slovakia in the years 1998-2018 based on multispectral satellite images from missions Landsat 5 and 7

Katarína Pukanská
Institute of Geodesy, Cartography and Geographical Information Systems, Faculty of Mining, Ecology, Process Control and Geotechnologies, Technical University of Košice, Košice, Slovakia

Faculty of Environmental, Geomatic and Energy Engineering, Kielce University of Technology, Kielce, Poland

Karol Bartoš, Juraj Gašinec & Ľubomír Kseňak
Institute of Geodesy, Cartography and Geographical Information Systems, Faculty of Mining, Ecology, Process Control and Geotechnologies, Technical University of Košice, Košice, Slovakia

Paweł Frąckiewicz
Faculty of Environmental, Geomatic and Energy Engineering, Kielce University of Technology, Kielce, Poland

Agnieszka Bieda
Department of Geomatics, Faculty of Mining Surveying and Environmental Engineering, AGH University of Science and Technology, Kraków, Poland

ABSTRACT: By using the instruments and methods of remote sensing, it is possible to detect forest damage caused by pests or calamities, while reducing time demands and achieving nationwide assessment of the area. This paper deals with the determination of the extent of deforestation in the Rožňava district (Slovakia) on the principle of calculating the normalized difference vegetation index NDVI from multispectral images of Landsat 5 and Landsat 7 satellites. The data sets were prepared in 5-year epochs using the JavaScript programming language and satellite bands B3, B4, B5, and B7 so that the cloudiness is eliminated from images while displaying the forest cover during the vegetation period. The NDVI calculation was performed in the ESA SNAP software. The resulting data classification was subsequently realized by calculating the percentage decrease in QGIS 2.18 software.

Keywords: Landsat 5, Landsat 7, multispectral images, NDVI, QGIS, deforestation

1 INTRODUCTION

1.1 *Type area*

Forest is one of the main components of the environment. The basic role of forests and forestry is to ensure long-term forest development, so they will continue to fulfil economic, environmental, and social functions in the future. Harmful factors cause significant problems in performing this task. These are of a natural character, specifically abiotic factors (windbreaks blew, windbreaks snow, windbreaks frost, damage by drought, frost damage, damage by floods, undetected causes of mortality in trees, freeze damage waterlogging) and biotic factors (bark and wood-destroying insects, fungi and animals) or anthropogenic character (direct and indirect human damage to forest stands). As a result, individual trees and stands perish. It may disrupt the non-production function of forests and causes great economic and environmental damage.

Information about forests have an areal natural character, and their definition is in space and time. To effectively process, manage, and analyse this data, it is necessary to implement, utilise, and develop modern tools such as geographical information systems (GIS) and remote sensing methods. To accurately quantify the change, we can accept conventional survey techniques that are slow and sometimes insufficient to make available information to planners promptly. With the introduction of remote sensing from aerospace platforms, a better way of collecting data is now available, providing an accurate, reliable and updated real-time database of land and water resources (Dash et al., 2018). Using satellite data, we can perform extensive forest stocktaking in the geographical area damaged by fire, using GIS tools and remote sensing methods to process and manage this data. Thus, remote sensing is a reliable method used to detect forested area and identify forest characteristics such as tree height, biomass and density [2].

1.2 *Study site*

The Rožňava district is located in the south-eastern part of Slovakia, covering an area of approximately 1173 kilometres square, while the highest peak named Stolica is 1476 meters high. Its southern border is identical with the state border with Hungary. The wood composition is mixed and consists of coniferous woods (30%) and deciduous woods (70%), while the complete wooded area is 723.1 square kilometres (61.6%). In terms of natural beauties and nature protection, there is an incredible number of exciting places. As for the Rožňava district, the most famous is National Park Slovak karst which is the largest karst area in Slovakia with many underground and surface karst formations. Well-known are caves, sinkholes, scrape fields with localization of important caves of tourism like Domica Cave, Gombasek Cave, and Dobšiná Ice Cave, which are part of the UNESCO cultural heritage (https://www.lesy.sk). As for forestry perspective, it is necessary to mention protected areas, national nature reserves such as Hrušovská forest steppe, Brzotín stones, and Zádiel gorge.

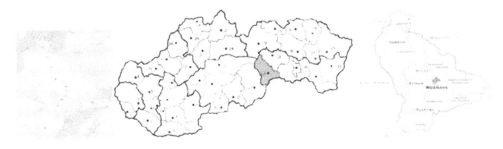

Figur 1.   Localization of Rožňava district.

The study area corresponds to part of one Landsat TM scene located in Slovakia in the region Rožňava – part of Gemer region, which has been affected by wind calamities and subcortical insects many times in this period. This region is presenting high deforestation rates and therefore has a high probability of forest degradation activities due to fire and selective logging.

According to the final reports of National Forest Centre (NLC) [3,4], the calamity of subcortical insects in spruce trees 1993 - 2012 caused that 23.1 million cubic meters of wood material damaged by bark and wood-destroying insects were processed in the period from 1993 to 2012 (20 years). However, up to 18.1 million cubic meters of wood has been processed in the last 10 years, which represents 78%. The calamity of subcortical and wood-destroying insects culminated in 2009, when almost 3.2 million cubic meters of wood mass. The most damaged regions were the spruce forests of Slovakia: the High Tatras, Low Tatras, Kysuce, Orava, Spiš and Gemer ("Filip" wind calamity of 23-24 August 2007). On 23 - 24 August 2007, a whirlwind swept through Slovakia. It was a southern wind that damaged forests in central Slovakia, especially in areas in Rimavská Sobota region. However, a large number of windbreaks (up to 700 000 cubic meters) were not processed at the end of 2007 for various reasons, especially in the branch of an enterprise Liptovský Hrádok.

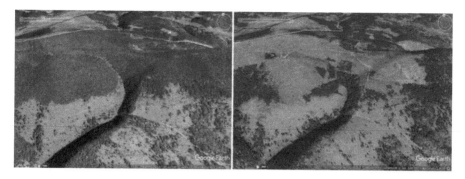

Figure 2.   Comparison of satellite pictures of the forest near to Rejdová village between 2006 and 2017.

2   DATA AND METHODS

2.1   *Remote sensing data*

For this research, we used multispectral images of Landsat 5 and 7 satellites as source data to calculate the area of deforestation (afforestation). Landsat 5 mission provided data from 3/1984 to 5/2013. The Landsat 5 satellite orbited the Earth in a sun-synchronous, near-polar orbit, at an altitude of 705 km, inclined at 98.2 degrees, and circled the Earth every 99 minutes. The satellite had a 16-day repeat cycle (Source: USGS.com). The Thematic Mapper (TM) is an advanced, multispectral scanning sensor. TM data are sensed in seven spectral bands simultaneously [5].

The government-owned Landsat 7 was successfully launched on April 15, 1999. The effect of unwanted pattern of Landsat 7 images was eliminated by the algorithm in Google Erath Engine. It is in a polar, sun-synchronous orbit with an altitude of 705 kilometres +/- 5 kilometres, it takes 232 orbits, or 16 days, inclination: 98.2°. Enhanced Thematic Mapper Plus (ETM+) products are delivered as 8-bit images with 256 grey levels.

2.2   *Data processing Google Earth Engine*

Earth Engine is a platform for scientific analysis and visualisation of geospatial datasets, for academic, non-profit, business and government users. Earth Engine hosts satellite imagery and stores it in a public data archive that includes historical earth images going back more than forty years. The images, ingested on a daily basis, are then made available for global-scale data mining (Source: [6]). Unlike Google Earth, which enables the discovery of the Earth through a virtual globe based on satellite imagery, maps, and terrain models, Earth Engine allows the analysis of water cover, afforestation, land-use and land cover change, and many other uses.

2.3   *Creating the JavaScript for download images*

Preparing the satellite imagery dataset in the vegetation period consisted of creating a JavaScript. Usually, satellite images can also contain clouds, which may present various problems in subsequent processing. In the Google Earth Engine platform, we chose the pre-prepared script, which we have customised according to our specific conditions - territory scope, time horizon. A script has been inserted into Earth Engine to obtain the images, which can be modified as needed. The script was written in JavaScript programming language, and the basic structure of the used code has already been created. The Earth Engine platform allows to upload and download scripts created by users. The next step was a further script modification for the final research purposes, i.e., adding the necessary commands and details.

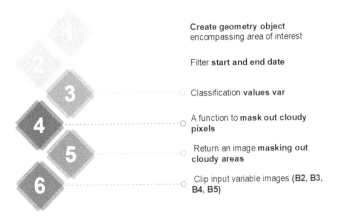

Create geometry object
encompassing area of interest

Filter **start and end date**

Classification **values var**

A function to **mask out cloudy pixels**

Return an image **masking out cloudy areas**

Clip input variable images (**B2, B3, B4, B5**)

Figure 3. WorkFlow of the script form Google engine [7].

The first part of the script contains a command defining the location coordinates which define the polygon of the area of interest. In the next part of the command, it was necessary to enter the name of the satellite whose images we want to use. Our analyses use images from Landsat 5 and Landsat 7 satellites. In the script for Landsat 5, the following satellite name was used: "LANDSAT/LT05/C01/T1_SR", and for Landsat 7: "LAND-SAT/LE07/C01/T1_SR". It was also necessary to specify the two dates of images execution. For this research, it was necessary to obtain images from the period when the forest is the densest, thus the most abundant vegetation. For this reason, summer months between 1 June and 31 August were chosen. In general, when the sensing interval is larger, the quality and quantity of images are better. Years of sensing period were changed for each image according to the purpose of the research, specifically years 1993 and 1998 for Landsat 5, and 2002, 2006, 2009, 2013 and 2018 for Landsat 7.

2.4   *Image processing in SNAP*

The generated images from the Earth Engine platform in *.tiff format were then imported into the ESA SNAP software. The processing continued with the calculation of the Normalized Difference Vegetation Index (NDVI index) (Figure 4). In SNAP, this calculation is possible thanks to the

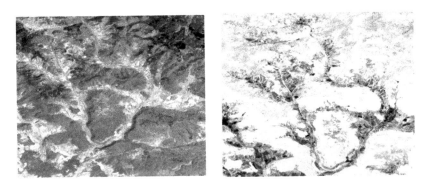

Figure 4.   Original image of the territory and NDVI image of Rožňava town and surroundings.

Batch processing application and using the "GraphBuilder" function, which allows starting the programmed process. NDVI is a simple graphical identifier used in remote sensing to detect vegetation from satellite images. It uses red (RED B3) and near-infrared radiation (NIR B4). Different ecosystem changes can be analysed from NDVI time series [1]. The next formula (1) represents the definition of NDVI calculation.

$$\text{NDVI} = \frac{X_{NIR} - X_{RED}}{X_{NIR} + X_{RED}} \tag{1}$$

2.5 Data classification

NDVI data classification was converted in SNAP software. Based on the specified forest area polygons, and subsequent statistical processing, the woodland area pixels were selected for each image. It means that forested pixels were associated with value 1, and non-forested pixels with value 0. This was the result of the so-called pixel mask of forest and non-forested areas (Figure 5).

2.6 Processing generated area and calculating of deforestation in QGIS

Further processing of generated areas was solved in the QGIS 2.18 programming environment using the MOLUSCE plugin (v.3.0.13). At first, it was opened the task window after running the plugin, where it was necessary to select two images in the "Inputs" tab. Specifically, it means the addition of the first image in the "Initial" field and allocation of the second image in the "Final" field. At the same time, both compared images were added to the "Spatial variables" field. After adding the compared images, the "Check geometry" function compared the image geometry for comparability. After the evaluation of image compatibility, all other bookmarks have been made available in the plugin. Calculation of the rate of deforestation has been carried out in the "Area Changes." Moreover, it is necessary to use the "Update tables" button in this bookmark. This function calculated the differences in the area of forested and deforested places between individual images and displayed the results in a percentage table. The result shows the percentage of forest area in both compared images, the percentage of deforestation in both images, and also the percentage difference between epochs. (Figure 6).

Figure 5.   Mask of forested and non-forested areas.

Figure 6.   Table of calculation of forested and non-forested areas.

Table 2.   Evaluation of the rate of deforestation between 2009 and
2013.

Area	1993	1998	Difference
Forested area	67,05 %	78,16 %	+11 %
Deforested area	32,94 %	21,82 %	
Area	1998	2002	Difference
Forested area	78,16 %	76,73 %	-1,44 %
Deforested area	21,82 %	23,27 %	
Area	2002	2006	Difference
Forested area	76,73 %	84,56 %	+7,83 %
Deforested area	23,27 %	15,44 %	
Area	2006	2009	Difference
Forested area	84,56 %	69,08 %	-15,48 %
Deforested area	15,44 %	30,91 %	
Area	2009	2013	Difference
Forested area	69,08 %	71,68 %	+2,59 %
Deforested area	30,91 %	28,32 %	
Area	2013	2018	Difference
Forested area	71,68 %	63,05 %	-8,62 %
Deforested area	21,82 %	23,27 %	
Area	1993	2018	Difference
Forested area	67,05 %	63,05 %	-4,00 %
Deforested area	32,94 %	36,94	

3   RESULTS AND DISCUSSION

Based on the processing, we obtained the following data on the rate of deforestation
(Table 2). The numerical values confirm the data from NLC databases [3], about the
extent of the calamity.

4   CONCLUSION

Remote sensing tools are currently an indispensable tool in monitoring environmental
changes. Using them, we can not only get real images but also process them and evaluate various analyses from them in a short period of time. The extent of the wind calamities and the
subsequent random extraction, which is evident in the Rožňava district, is confirmed by satellite images and figures.

REFERENCES

[1] Forkel, M., Carvalhais, N., Verbesselt, J., Mahecha, M.D., Neigh, Ch., Reichstein, M. 2013. Trend Change Detection in NDVI Time Series: Effects of Inter-Annual Variability and Methodology. *Remote Sensing, MDPI, Basel, Switzerland, 2013*, 5(5): 2113–2144.

[2] Dong, G., Yang, J., Peng, Y., Wang, Ch., Zhang, H. Forest characteristic detection with Pol-SAR, *Journal of Tsinghua University (Science and Technology)*: 2003–07.

[3] Kunca, A., Galko, J., Zúbrik, M. 2014. Významné kalamity v lesoch Slovenska za posledných 50 rokov. In: Kunca, A. (ed.): Aktuálne problémy v ochrane lesa, *Národné lesnícke centrum, Zvolen, 2014*: 25–31.

[4] Kunca A., Zúbrik M.: Výskyt škodlivých činiteľov v lesoch Slovenska v rokoch 1960-2014 a v roku 2015 a prognóza ich vývoja, Národné lesnícke centrum - Lesnícky výskumný ústav Zvolen, 2016, str.139, ISBN 978-80-8093-219 – 0.

[5] USGS.com. Landsat Missions - Landsat 5 mission characteristic. https://www.usgs.gov/land-resources/nli/landsat/landsat-5?qt-science_support_page_related_con=0#qt-science_support_page_related_con, (Available on the Internet).

[6] EarthEngine.com. Google Earth Engine platform. https://earthengine.google.com/faq/(Available on the internet).

[7] https://developers.google.com/earth-engine/datasets/tags

Advances and Trends in Geodesy, Cartography and Geoinformatics II –
Molčíková, Hurčíková, & Blišťan (eds)
© 2020 Taylor & Francis Group, London, ISBN 978-0-367-34651-5

Integration of multispectral and radar satellite images for flood range determination of historical flood in Slovakia 2010

Katarína Pukanská

Institute of Geodesy, Cartography and Geographical Information Systems, Faculty of Mining, Ecology, Process Control and Geotechnologies, Technical University of Košice, Košice, Slovensko

Faculty of Environmental, Geomatic and Energy Engineering, Kielce University of Technology, Kielce, Poland

Karol Bartoš, Viera Hurčíková & Ľubomír Kseňak

Institute of Geodesy, Cartography and Geographical Information Systems, Faculty of Mining, Ecology, Process Control and Geotechnologies, Technical University of Košice, Košice, Slovensko

Marcin Gil

Faculty of Environmental, Geomatic and Energy Engineering, Kielce Univerity of Technology, Kielce, Poland

Agnieszka Bieda

Department of Geomatics, Faculty of Mining Surveying and Environmental Engineering, AGH University of Science and Technology, Kraków, Poland

ABSTRACT: The objective of this work is to evaluate the extent of historic floods on rivers within Košice region in June 2010. Nowadays, massive floods are an annual reality, as the increasing number of natural disasters is a consequence of global climate changes. Their increasing frequency and severity cause immense damage to property, but also loss of human lives. For the purpose of mapping these floods, historical multispectral satellite images from Landsat 5 were used as the input data, together with radar image from Envisat, from the time of floods in 2010. The results of the floods were compared with the corresponding satellite data from the data with a normal situation. All used image and radar data are freely available to download and process. Given data sets were opened and processed in open-source software ESA-SNAP. The result of the study is the determination of the extent and quantification of historical floods in Eastern Slovakia in 2010.

1 INTRODUCTION

1.1 *Floods in Slovakia in the year 2010*

Flood is a natural process, during which the water will temporarily flood the usually non-flooded area [1]. Floods are mainly phenomena at which the level of the watercourse reaches a level that can be considered socially and/or economically dangerous. Floods are a permanent part of the water cycle in nature, and are an extreme hydrological phenomenon that has occurred in the past, is occurring now and will continue to occur in the future. In 2010, the east of Slovakia was hit by floods since mid-May. The upper parts of the Hornád River and its tributaries were hit by floods with moderate significance, while the middle and lower parts, as well as the watercourses in southeast Slovakia, were hit significantly with catastrophic consequences in this year. At the beginning of June, extreme rainfall definitely hit the water-saturated basin and caused floods with

50-100-year significance in many places [2]. The historical importance of these rainfalls is also confirmed by the fact that the so far valid records of monthly rainfall totals in May have been broken at almost 400 rainfall stations in Slovakia (many of these rainfall stations record monthly rainfall totals since 1901).

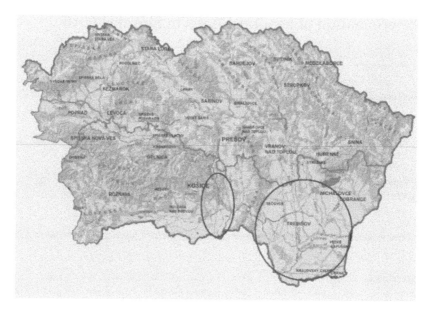

Figure 1. Eastern Slovakia - areas affected by floods.

After evaluating the hydrological situation, weather situation and consequently runoff situation in the catchment areas of eastern Slovakia, the Department of Hydrological Forecasts and Alerts Košice issued continuously updated 1st, 2nd and 3rd degree hydrological alerts to floods from permanent rains and storm floods from 6.5.2010, and was monitoring and informing by sending extra hydrological information to the relevant institutions.

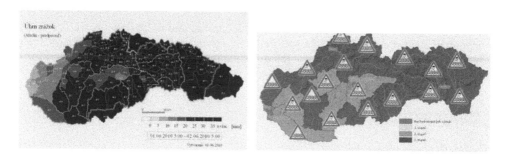

Figure 2. Total rainfall during 1.6.-2.6.2010 and flood warning of 3.6.2010 at 10.00 [2].

The flood situation in May and June 2010 on eastern Slovakia's watercourses has caused significant damage to property such as landslides, damaged family houses, gardens, cottages, damaged state roads and local roads, bridges, cemeteries, flooded buildings, wells and outbreak of mosquitoes after floods. In addition to material damage, this flood has also claimed human lives. From 6.5. to 11.6., 5378 persons were evacuated in the Košice region. The flood has caused significant changes in many watercourses, such as clogging or deepening of the channel [2].

2   MATERIALS AND METHODS

2.1   *Data source*

Several historical satellite data sources were used to calculate the extent of the floods. In 2010, several multispectral and radar satellites were available. For this research, we focused on freely available, non-commercial multispectral data from the NASA Thematic Mapper sensor of Landsat 5 (L5) satellite; and the radar data from the ESA Envisat satellite (now non-functional) which were needed mainly due to the penetration of radar radiation through the clouds during rainfall – phenomena accompanying floods.

The Thematic Mapper (TM) is an advanced, multispectral scanning sensor. TM data are sensed in seven spectral bands simultaneously. Ground Sampling Interval (pixel size): 30 m reflective, 120 m thermal Temporal resolution of the sensor is 16 days, image size is 185 x 172 km and altitude is 185 km.

The ESA Envisat satellite was the world's largest civilian Earth observation satellite on the Sun-synchronous polar orbit. Repeat interval of the satellite was 35 days with altitude 790 km ±10 km with inclination 98,40°. SAR (Advanced Synthetic Aperture Radar) operates in the C band in a wide variety of modes.

For the comparison and evaluation of floods, we processed images with the normal situation from 24 April 2010 (L5) and from 20 September 2010 (Envisat); and images during floods from 5 June 2010 (L5) and from 7 June 2010 (Envisat) (Figure 3).

2.2   *Processing of satellite data in ESA SNAP and QGIS*

The European Space Agency (ESA) SNAP - Sentinel application platform was used for satellite data processing. The essence of the data evaluation was the processing of the image by appropriate methodical procedures to obtain the masks of flooded areas. For processing multispectral images of L5, the Normalized Difference Water Index NDWI, or Modification of Normalized Difference Water Index MNDWI, can be used. Using both, we can measure the extent of the surface water and flooded areas. It was first proposed by McFeeters in 1996 [3]. It is a satellite-derived index from the Green (G), Near-Infrared (NIR) and Middle infrared channels (MIR).

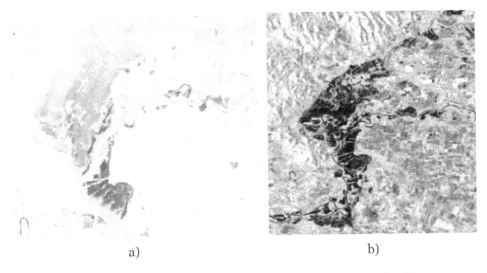

a)   b)

Figure 3.   Original satellite image of the flooded area from Landsat 5 (a), and Envisat (b).

NDWI formula [3]:

$$NDWI = \frac{X_{green} - X_{nir}}{X_{green} + X_{nir}},\tag{1}$$

This index is designed to maximise water reflectance using green wavelengths, minimize low NIR reflectance by water features, and utilise high NIR reflectance by vegetation and soil. As a result, water bodies have positive values and are better suited for processing. On the contrary, vegetation and soils usually have zero or negative values and are therefore suppressed.

MNDWI formula [4]:

$$NDWI = \frac{X_{green} - X_{mir}}{X_{green} + X_{mir}},\tag{2}$$

By calculating MNDWI, we can get more positive values than NDWI because MIR absorbs more light than NIR. The built-up area will have negative values; soil reflects MIR more than NIR, vegetation reflects MIR even more than GREEN. Compared to NDWI, the contrast between water and built-up land will therefore significantly increase due to increasing watercourse values and decreasing built-up area values from positive to negative.

The Envisat SAR data were processed by batch processing by applying the following steps (Figure 4):

Figure 4.   Batch processing of Envisat SAR data.

Read: Loading the data being processed. **Apply Orbit File:** The orbit state vectors provided in the metadata of a SAR product are generally not accurate and can be refined with the precise orbit files. The orbit file provides accurate satellite position and velocity information. **Calibration:** The objective of SAR calibration is to provide imagery in which the pixel values can be directly related to the radar backscatter of the scene. Though uncalibrated SAR imagery is sufficient for qualitative use, calibrated SAR images are essential to the quantitative use of SAR data. **Speckle filtering:** For polarimetric SAR data, the speckle filtering is based on incoherent averaging and requires handling statistical second-order representations (used filter Lee Sigma, window size 7 x 7). Thus, the speckle filtering is applied to covariance or coherency matrix. **Terrain correction:** Due to topographical variations of a scene and the tilt of the satellite sensor, distances can be distorted in the SAR images. Image data not directly at the sensor's Nadir location will have some distortion. Terrain corrections are intended to compensate for these distortions so that the geometric representation of the image will be as close as possible to the real world (Figure 5). SRTM 3Sec digital elevation model was used. **Write**: Write image.

The geometry of topographical distortions in SAR imagery is shown below. Here we can see that point B with elevation h above the ellipsoid is imaged at position B' in SAR image, though its real position is B". The offset Δr between B' and B"exhibits the effect of topographic distortions.

2.3   *Creating a mask with the flooded area*

To create the mask with flooded area from both prepared images L5 and Envisat, the area of water region (a) must be carefully selected, and divided it into two groups with the values of 0 for the non-flooded areas and the value of 1 for the flooded area, based on statistical analysis of pixel intensities in images (b). In the case of the MNDWI image, the flooded area was identified for intensity values 1 < pix <0.90. Similarly, with the region selection and statistical analysis, we proceeded with radar data - 1 < pix <0.90 (Figures 6 and 7).

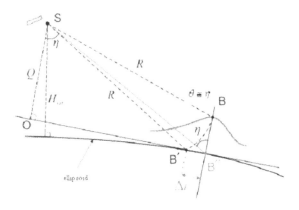

Figure 5.   Terrain correction on SAR image [5].

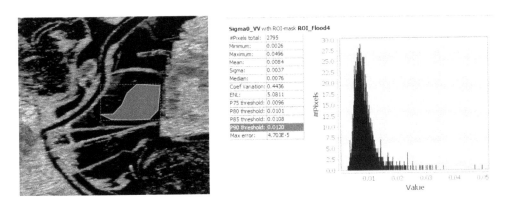

Figure 6.   Selection of the flooded region by a polygon, and statistical calculation of pixel intensity on the processed radar image.

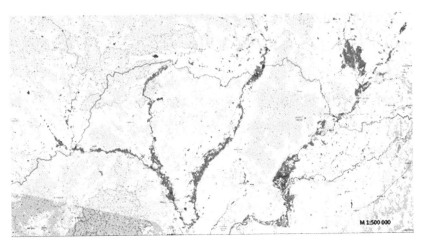

Figure 7.   Image of the mask of the original and flooded watercourses of the rivers Hornád, Torysa, Bodrog, and Tisza, in the territory of the Slovak Republic and Hungary, at a scale of 1:500 000.

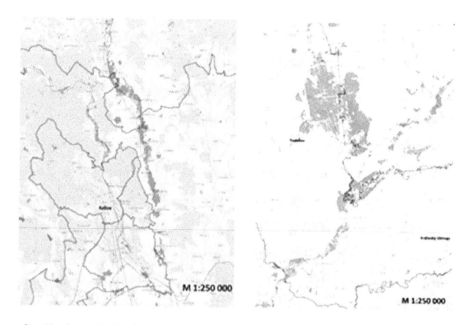

Figure 8. Flood maps in the district of Košice and Košice-okolie, and Trebišov, at a scale of 1:250 000.

Table 1. Resulting flooded areas for individual rivers.

River catchment	Landsat 5 (5.6.2010)	Envisat (7.6.2010)
Torysa	30,9 km^2	-
Hornád	34,2 km^2	-
Ondava/Bodrog	142,9 km^2	130 km^2
Latorica	32,2 km^2	15 km^2

From the processed images, after calculation, the flooded area in the districts of Košice, Košice – okolie, Prešov, Trebišov, and Michalovce was determined (Figure 8).

The resulting values of the flooded areas on the Torysa, Hornád, Ondava, Bodrog and Latorica were determined as follows (Table 1).

3 CONCLUSION

Remote sensing tools provide great opportunities to quickly and accurately monitor changes on the Earth's surface. The constantly evolving technologically new sensors with greater spatial, radiometric and temporal resolution can now replace other time-consuming measuring technologies.

ACKNOWLEDGEMENTS

The study is the result of Grant Project of Ministry of Education of the Slovak Republic KEGA No. 004TUKE-4/2019: "Scientific and educational centre for remote sensing with the focus on the application of e-learning approaches in education".

REFERENCES

[1] Ochrana pred povodňami, Ministerstvo životného prostredia SR, https://www.minzp.sk/oblasti/voda/ochrana-pred-povodnami/

[2] Slovenský hydrometerologický ústav SR, Povodňová situácia na tokoch východného Slovenska v roku 2010. http://www.shmu.sk/File/Povodnova_situacia_na_tokoch_vychodneho_Slovenska_v_maji_a_juni_2010_opravene1.pdf

[3] McFeeters, S.K. The use of the Normalized Difference Water Index (NDWI) in the delineation of open water features. *Int. J. Remote Sens* 1996, 17, 1425–1432.

[4] Hanqiu XU, Modification of normalised difference water index (NDWI) to enhance open water features in remotely sensed imagery, International Journal of Remote Sensing Vol. 27, No. 14, 20 July 2006, 3025–3033, http://www.aari.ru/docs/pub/060804/xuh06.pdf.

[5] SNAP. (2019). ESA.

Advances and Trends in Geodesy, Cartography and Geoinformatics II –
Molčíková, Hurčíková, & Blišťan (eds)
© 2020 Taylor & Francis Group, London, ISBN 978-0-367-34651-5

Deformation measurement of the Orlík dam – evaluation and innovation of methods

J. Seidl, T. Jiřikovský & M. Štroner
Department of Special Geodesy, Faculty of Civil Engineering, CTU in Prague, Czech Republic

ABSTRACT: The Orlík Water Reservoir is located in the valley of the Vltava River, 91 km south of Prague and is the largest dam in the Czech Republic. The main purpose of the dam is to produce electricity, accumulate water to improve flow rates at the bottom of the Vltava, partially protect the area below the dam and also Prague from large waters. Today, the dam is one of the largest recreational and sports areas in the Czech Republic. It was built in years 1954 – 1960, the overall volume is more than 700 mil cubic meters. The dam is built from reinforced concrete and is 450 m long a more than 90 m high. The article presents the reference network stabilized by the concrete pillars, observed points, used monitoring methods, the measurement procedure and the determined shifts for the last 4 stages, both reference network and monitored points. Results are accompanied by the accuracy analysis. Article also includes a proposal to complement the present measurement and innovation of existing methods, i.e. replacement of used methods by modern ones, made on the base of experiences acquired during the practical measurement.

1 INTRODUCTION

Measurement of shifts and deformations of water constructions is a very important part of geodetic work, the main reason is to ensure safety, especially early detection of building structure defects. The monitoring is carried out by various methods, and modern methods of mass data collection allowing the area monitoring as well as automatic methods enabling essentially continuous monitoring without the presence of the operator/meter are undergoing great development. The development of geodetic methods for monitoring changes in dams is addressed by a number of work, with the possibility to mention e.g. [1,2]. Remote sensing methods ([3-5]), global positioning systems [6] as well as computational methods [7] are also developing. However, it is not possible to forget the traditional methods (innovated e.g. in [8]), which have an unquestionable significance especially in older and already established water structures. These methods were principally based on historical instruments (from today's point of view), which need to be gradually replaced and, in conjunction with this, the methods used can also be modernized, but still for monitoring individual characteristic points. The article describes the ways of monitoring on water works, the specific solution of monitoring the dam of the Orlík reservoir in the Czech Republic, including a description of the geodetic network and observed points, used surveying methods and evaluation of displacements of pillars of the survey network and observed points in the last 4 stages. At the end, there are given the results of the accuracy models, which form the basis for the design of the current survey network and the innovation of measurement methods.

2 WATER DAM ORLIK

The Orlík reservoir is the largest one in the Czech Republic with a reservoir volume of approximately 720 million cubic meters of water and a bank length of over 300 km. The dam

Figure 1. The Orlík water dam.

itself is made of reinforced concrete and its construction is belonging among the heavy dams. In the crown, it reaches a length of 450 m and a height of over 90 m and is divided by expansion joints into 33 blocks. There are 3 crown spillways in the dam and a hydroelectric power station is located under the dam (Figure 1). The main purpose nowadays is water retention, flow improvement at the downstream of the Vltava River, electricity production, behind the dam is a water reservoir, which is one of the largest recreational and sports areas in the Czech Republic.

3 PRINCIPLES OF SAFETY MONITORING OF LARGE DAMS IN CZECH REPUBLIC

In general, there is no clear procedure for monitoring safety of the dams in the Czech Republic legislation, but specific methods are chosen for each water work, considering its shape, size and conditions not only in the surrounding area but also within the work itself. Binding procedures and methods are subsequently described in the Technical-Safety Supervision Program, which is prepared by the relevant administrator and the head of technical and safety supervision on the work. Three basic parameters are being followed in practice to ensure safety. Horizontal dam shifts, vertical dam shifts and dam deflection.

In order to monitor vertical displacements, the method of very precise or precise geometrical leveling is chosen in most cases due to high accuracy, which, in addition to shifts of the dam itself, also verifies the stability of fixed points in the surroundings and possibly shifts of the basin and shores.

In practice, the monitoring of horizontal displacements is divided into two aspects, namely the shift in the "direction" of the flow and the "direction perpendicular" to the flow. For both directions of displacement, the method of the alignment is most often used, i.e. the determination of the deviation angle changes, which is still based on the time when it was easier to determine the angle than the distance. The alignment can be done in two forms, either one-sided, where the deviation angle is determined from one viewpoint with orientation and network points, or double-sided, where the deviation angle is determined for two views and the resulting shift is determined by a weighted average. Both methods require verification of the stability of observation and orientation points. This is done using either network points where a minimal change of position is assumed or by building very accurate surveying network, which is measured and then adjusted in each epoch. Pendulum systems are most commonly used to monitor tilts and deflections, whether with manual or automatic readings at the required levels. Other methods may be the use of clinometric and inclinator bases. The above-mentioned methods are used to determine absolute displacements, physical methods such as dilatometers and strain gauges are used to monitor relative displacements such as expansion joints.

4 RECENT STATE AND MEASUREMENTS ON ORLIK DAM

Since the Orlík reservoir was built in the 1950s, almost all methods for monitoring safety can be found here. For vertical displacements, the method of precise leveling is used here, both for the measurement of displacements outside the dam and for monitoring in revision galleries. Horizontal displacements in the direction of flow are monitored by means of a double-sided alignment method, where 24 double-sided enamel targets are installed on the dam, and were measured from the the pillars of the original surveying network, which consisted of 9 pillars in total.

In 1995, the existing network was expanded by three new pillars, which were in better configuration for the alignment method, and therefore monitoring switched to two new pillars with stability verification by the network adjustment. The geodetic network was no longer fully measured because of economic reasons, but only the part of realized by the seven pillars (Figure 2).

Figure 2.   Part of reference network (maps.google.com).

The horizontal shifts perpendicular to the flow are measured from the original network points and the one-sided alignment method is used with verification by the adjustment. Eight Huggenberger dam pendulums are used for tilt and deflection monitoring.

5 ACTUAL EPOCHS AND THE TEST MEASUREMENT

The article considers measurement from the last 4 stages, which were carried out in cooperation with the company Vodní stavby - TBD a.s., which is in charge of technical safety supervision of the Orlík waterworks. The basic stage was measured at the beginning of August 2018, followed by another in October and December and the fourth stage in March 2019.

During the basic and third stage, the horizontal displacements on the air face of the dam were measured only during the period of maximum pendulum deflections. At all stages, the surveying network was measured to verify its stability and also the observed points on the dam body in the form of Leica mini-prisms, which were installed here in the basic stage and serve to test of the usability of the polar method for of the determining of shifts and deformations of the dam.

As mentioned above, a double-sided alignment method is used to monitor horizontal displacements, which is measured from two concrete pillars whose line is approximately parallel to the dam body. In addition to considering atmospheric effects on measurements, it is also necessary to consider the eccentricities of individual observation points from the baseline, which points are verified by the measurements with use of the geodetic network. In order to ensure the highest accuracy of the network, the concrete pillars are equipped with a system of forced centering, and during the measurement, each pillar has a clearly defined assembly (consisting of a tripod and prism) and its position and rotation to limit the systematic errors between stages (make them equal).

During the measurement, the influence of atmospheric conditions on the measurement has to be considered, the important factor influencing the accuracy is the refraction, whose vertical component has a great influence on polar measured variables, especially due to the height differences in the network and differently irradiated lines of sight.

The effect of the refraction can then be monitored in the results of the accuracy analysis, which must be performed before the measurements are adjusted to determine the accuracy of the measured variables within the network and determine the appropriate standard deviation values for the weight calculation. The adjustment takes place in the form of the free network with subsequent Helmert transformation to the center of gravity of the network, which is the same for all stages to avoid misinterpretation of shifts due to the shift of center of gravity between stages.

The resulting points shifts are then determined from the difference of the adjusted coordinates of the survey pillars, and their provability is tested by the criterion calculated as the standard deviation of the shift multiplied by the coefficient $u_p = 2.5$ (corresponds to significance level $\alpha = 0.01$, normal probability distribution, according to [9]).

Similarly, the shifts of observed points are also tested, where the magnitude of the interstage shift is converted from the difference of the deviation angles to the magnitude by the arc-to-arc conversion and the standard deviation of the displacement is determined by law accumulating the directional deviations of the formula for the two-way intention line. Similarly, the shift of observed points is also tested, the standard deviation of the displacement is determined by the standard deviation propagation law applied on the formula for the two-sided alignment. The shifts determined between the stages are shown in Figure 3 and Figure 4.

6 ACCURACY MODELS

Prior to the measurement, a priori precision models were created for both the network and the observed points. The use of precision models was then the basis for the analysis of the design of the alignment method replacement by the spatial polar method and the proposal of the extension of the reference network. For the method innovation, the approximate coordinates of the observed points were determined and subsequently the measurement from the P7 pillar, which is

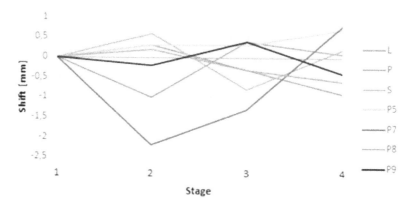

Figure 3.   Shifts (in direction of flow) of the survey network points.

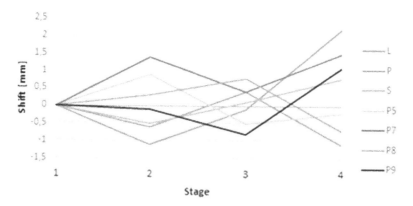

Figure 4.   Shift (in direction perpendicular flow) of the survey network points.

located approximately in the dam axis, was determined. Special position of the pillar P7 allows to measure all the dam sections with practically identical standard deviation of all adjusted coordinates.

In addition to the comparison of theoretical accuracy, there was also a made a comparison of the real achieved accuracy of the alignment method with the accuracy of the adjusted coordinates of the Leica miniprisms.

The extension of the reference network is planned due to the construction of a safety spill slip on the right bank of the river, during which at least one existing pillar of the surveying network will be destroyed and the construction of three new ones is expected. As the slip body is located outside the area of the currently measured network, a model was created, which in addition to the existing surveying network and new pillars, is also considering the use of a part of the old, now unused surveying network. The main reason for it is to ensure greater reliability within the network for both construction and future monitoring. An important aspect here is also the economic factor, because with increasing number of pillars there is a longer measurement time but also an increase in the costs associated. The results of modeling are shown in Table 1, where the increment of the accuracy in the reference network can be seen when two pillars are added.

Table 1. Precision modelling results.

Current network			Current network with P3 and P6		
Pillar	s(x)/mm	s(y)/mm	Pillar	s(x)/mm	s(y)/mm
L	0.26	0.22	L	0.24	0.18
S	0.11	0.11	S	0.08	0.07
P3	—	—	P3	0.16	0.16
P5	0.19	0.16	P5	0.15	0.14
P6	—	—	P6	0.09	0.09
P7	0.12	0.11	P7	0.08	0.07
P8	0.21	0.16	P8	0.17	0.14
P9	0.13	0.11	P9	0.08	0.06
P-new	0.14	0.15	P-new	0.10	0.12
S1	0.11	0.10	S1	0.07	0.06
S2	0.13	0.13	S2	0.09	0.08

7 CONCLUSIONS

The horizontal displacements of the observed points were demonstrated in virtually all perineum blocks (with a 1% risk of error). The main reason for this is, among other things, the thermal expansion of the concrete and the changing amount of water in the dam.

Interesting is the comparison of three dam blocks where the observed points are monitored simultaneously by the polar method using automatic targeting and the alignment method, where the results differ in the order of tenths of a millimeter, both can be considered to be correct and within the accuracy of both methods. Within the independent processing of the last 4 stages, it is possible to trace the instability of some pillars, mainly the pillar P7, which is located directly on the body of the water lift and is heavily influenced by the thermal expansion.

For the remaining pillars, there was always at least one stage of proven shift with 1% risk. In addition to the movement of the pillar itself, it is necessary to consider the influence of the geological subsoil and its periodic movements. However, since only four stages were evaluated, it is not possible to consider possible movements of subsoil and actual shifts of individual pillars and therefore a clear verification of the stability of the measuring network.

During verifying the idea of replacing the alignment method by the polar method, there was an interesting finding for the Orlík dam, theoretical analysis showed 50% better accuracy for existing methods than for new methods, but this difference is significantly reduced when using realistically achieved accuracies (real accuracy of measurement, accuracy of targeting).

Since the achievable accuracy is fit for purpose, it is possible to consider replacing the alignment method with a spatial polar method, as this saves time and reduces measurement effort. Due to the results of the verification measurement, it is recommended to fit additional blocks of the dam by mini-prisms for further testing during a longer time period.

From the accuracy models for extending the existing network, it can be concluded that against the existing measurement configuration will be reduce the internal reliability of the geodetic network due to construction. For this reason, another theoretical analysis was performed by testing results when adding additional points (pillars with forced centering). The most efficient model was with added 2 points – on - pillars, where internal accuracy achieves similar results to the current configuration. By adding more pillars, the resulting accuracy can be increased further, but it is also necessary to consider the increasing cost of maintenance and measurement of the geodetic network (including higher time demands).

REFERENCES

[1] Lienhart, W.: State-of-the Art of Geotechnical Monitoring with Geodetic Techniques. GEOTECH-NICAL ENGINEERING, 50, 2, 2019, ISSN: 0046-5828.

[2] Kuras, P. - Ortyl, L. - Owerko, T. - Borecka, A.: Geodetic monitoring of earth - filled flood embankment subjected to variable loads. REPORTS ON GEODESY AND GEOINFORMATICS, 106, 1, 2018. DOI: 10.2478/rgg-2018-0009. ISSN: 2391-8365.

[3] Scaioni, M. - Marsella, M. - Crosetto, M. - Tornatore, V. - Wang, J.: Geodetic and Remote-Sensing Sensors for Dam Deformation Monitoring. SENSORS, 18, 11, 2018, DOI: 10.3390/s18113682, ISSN: 1424-8220.

[4] Corsetti, M. - Fossati, F. - Manunta, M. - Marsella, M.: Advanced SBAS-DInSAR Technique for Controlling Large Civil Infrastructures: An Application to the Genzano di Lucania Dam. SENSORS, 18, 7, 2018, DOI: 10.3390/s18072371, ISSN: 1424-8220.

[5] Voege, M. - Frauenfelder, R. - Larsen, Y.: DISPLACEMENT MONITORING AT SVARTEVATN DAM WITH INTERFEROMETRIC SAR. In: IEEE International Symposium on Geoscience and Remote Sensing IGARSS. 2012, Munich, Germany. ISBN: 978-1-4673-1159-5.

[6] Manake, A. - Kulkarni, M. N.: Study of the deformation of Koyna Dam using the Global Positioning System. SURVEY REVIEW, 36, 285, 2002, ISSN: 0039-6265.

[7] Nowel, K.: Robust M-Estimation in Analysis of Control Network Deformations: Classical and New Method. JOURNAL OF SURVEYING ENGINEERING, 141, 4, 2015, DOI: 10.1061/(ASCE) SU.1943-5428.0000144, ISSN: 0733-9453.

[8] Macháček, T. - Krnáč, V.: Using high-precision total station operating in ATR mode and robust adjustment of geodetic networks for safety supervision over waterworks. In: International Symposium on Water Management and Hydraulic Engineering, Brno, Czech Republic, 2015. ISBN: 978-80-214-5230-5.

[9] Hampacher, M. - Štroner, M.: Zpracování a analýza měření v inženýrské geodézii. 2. vyd. Praha: Česká technika - nakladatelství ČVUT, ČVUT v Praze, 2015. 336 s. ISBN 978-80-01-05843-5.

Advances and Trends in Geodesy, Cartography and Geoinformatics II –
Molčíková, Hurčíková, & Blišťan (eds)
© 2020 Taylor & Francis Group, London, ISBN 978-0-367-34651-5

The accuracy determination of landslide cone terrain mapping by the UAV photogrammetry in High Tatras

M. Štroner & R. Urban
Department of Special Geodesy, Faculty of Civil Engineering, CTU in Prague, Czech Republic

P. Blišťan, L. Kovanič & M. Patera
Institute of Geodesy, Cartography and Geographical Information Systems, Faculty of Mining, Ecology, Process Control and Geotechnology, TU Košice, Slovak Republic

ABSTRACT: Mass data collection technologies are widely used today and allow very detailed capturing of virtually any objects, including the Earth's surface, for example for monitoring changes. Increasingly, UAV photogrammetry is used not only in geodesy for this purpose, it provides significant advantages in the mapping of complex landslide-type objects compared to terrestrial surveying methods, including 3D scanning. The paper presents a test of the precision of the mapping of the surface of the stonefield landslide in the Little Cold Valley in the High Tatras made by UAV photogrammetry by comparing its results with the point cloud acquired by the Leica P40 terrestrial 3D scanner. The evaluation not only considers the accuracy of the result, but also compares the cost-effectiveness of the measurement as well as the cost of acquiring the necessary equipment. Measured terrain is very specific, rugged, it is formed by boulders of the size of meters over smaller stones to gravel and sand. The slope of the terrain is also extreme. Evaluating the cost-effectiveness and meticulousness of the measurement will stand out more than in normal situations, as the measurement took place in a national park where all equipment, devices and personal baggage need to be transported by human power. At the same time, the movement in the stone field is even more difficult. It has been found that the methods compared are practically equal in terms of precision, but drone photogrammetry is clearly better in terms of labor and economy.

1 INTRODUCTION

The measurement of shifts and deformations is common in the field of geodesy and cartography, where the discipline is technically and practically highly sophisticated, which uses especially terrestrial measurement by total stations and monitoring of individual signaled points with high accuracy at the level of millimeters. These procedures are capable of internal accuracy analysis allowing the determination and control of the accuracy achieved, but it is not suitable for detailed monitoring of area or spatially irregular formations, as signaling and subsequent measurement of points is unbearable in terms of time and money. Therefore, mass data collection methods are used for these monitoring, which include mainly terrestrial and aerial 3D scanning and photogrammetric methods based mainly on the Structure from Moving (SfM) method, which is significantly more cost-effective than 3D scanning, whether terrestrial or from air means.

Because photogrammetric (or 3D scanning) methods do not allow in principle to be placed in a global coordinate system (georeferencing), it requires determination of a certain number of ground control points using terrestrial geodesy (with a global system connection) or more often GNSS (global navigation satellite system, with subsequent transformation) into the terrestrial position and height system. If ground control points cannot be considered stable and unchanging, which is often not possible, the highest achievable accuracy is determined by the accuracy of their determination. There are many cases in the literature for determining changes in natural formations, landslides, etc. In the case of [1-9], the presented cases illustrate the measurement principles and procedures generalized in the paragraph above, and at the same time document a commonly achievable accuracy of ten to five centimeters. Monitored terrain is usually smooth, practically free, or with very thin vegetation. Another and less frequently solved problem is the monitoring of areas covered by vegetation, which makes it difficult to determine changes.

The main objective presented in this paper was to test methods suitable for the monitoring of landslides in specific high-altitude conditions, as well as the procedure for monitoring the debris cone landslides using the SfM method implemented by UAV imaging. A significant problem is the nature of the terrain, which consists of practically only larger or smaller boulders, and it is therefore necessary to take pictures so that a continuous surface can be reconstructed from the data. Due to the terrain segmentation, it is virtually impossible to control surface capture by other means than by mass data collection, if possible more accurate and reliable. To do this, 3D scanning was chosen. The result of the work is the evaluation of quality, laboriousness and further usability of the tested measurement methodology for the needs of the slope landslide monitoring with regard to the landslide risks in the very specific terrains of the High Tatras.

2 TEST SITE

The Small Cold Valley (Malá Studená dolina) was chosen as a testing site due to the nature of the terrain, high slope and relatively good accessibility. There are manifestations of massive rubble stream under the walls of the Lomnický Peak, whose source area starts at an altitude of approximately 2,200 m, its length in 2014 was approximately 1,250 m, ie 50 m more than in 2004. Its width also was increased from the original 5 to 12 m to 96 m in the storage zone. An illustration of the area is shown in Figure 1 and Figure 3.

Figure 1.   Geodetic network on the site (square - point, triangle – temporary standpoint; map - maps. google.com).

3   USED INSTRUMENTS

For long-term monitoring of the tallus cone by methods of mass data collection (photogram-metry and laser scanning) it was necessary to build a geodetic network in order to evaluate changes in individual stages. The grid point stabilization was primarily made by copper studs into the rock using a chemical anchor. For effective measurement, the spot field was further supplemented with several reflective labels that were glued to suitable flat rocky surfaces or large rocks in the valley using a special glue. There were also three temporary points using GNSS (RTK method connected to the Leica SmartNet network), which formally used to con-nect a geodetic network to JTSK and Temporary GNSS points were used only to determine the spatial position and orientation of the total station, and then everything was measured by the spatial polar method (point array, UAV ground control points, scanner ground control points). Geodetic measurements were performed with a total station Leica TS02 with an angu-lar measurement accuracy of 7" (0.0020 gon) and a distance measurement accuracy of 2 mm + 2 ppm d. During the measurement, the spatial position of the total station was checked regu-larly. The drone DJI Phantom 4 Pro was used to capture the stone field and has a 5472 x 3648 pixel camera. Altogether, 1389 images were taken in several air raids from an average height of 35 m above ground, which means that 1 pixel represents 0.01 m in reality. During the flight, automatic camera settings were used (ISO 100, Shutter 1/60 to 1/800, F/3.5 - F/7.1). Total flight time was about 3 hours. Ground control points for UAV were made of fiberboard 0.3 x 0.3 m with black and white target.

Laser scanning was performed with the Leica P40, which features a two-axis compen-sator, 1.2 mm + 10 ppm distance measurement accuracy, 8 "(0.0025 gon) angle measure-ment precision, in 3D 3mm/50m and 6mm/100m point accuracy, scanning speed up to 1 million points per second and 360 ° x 270 ° field of view. For each scanner standpoint, at least three temporary ground control points (black and white Leica targets) were measured by a total station to register the resulting clouds. The scanner resolution was set to 12mm/10m within a range of 120 m. Altogether, 25 positions with a duration of about 12 hours of measurements were used because the scanner carrying in the protective box is weighing about 30 kg in a stone field was very problematic. Also, worth mention-ing the impossibility of mechanized transport in the national park, the equipment and all the equipment had to be transported to the place of measurement by human forces, instruments were largely taken up by a member of the team of authors Matej Patera, otherwise a voluntary Tatra carrier.

4   DATA PROCESSING

3D scan data was processed only by transforming individual scans into a common coordinate system defined by the stabilized points of the geodetic network, with which individual ground control points were targeted by the total station. Further processing was not necessary, the resulting cloud contained 597 million points. The processing was carried out in the Leica Cyc-lone program, the resulting data being shown in Figure 2. At the top right, you can see the measurement from one standposition of the instrument, where due to the nature of the terrain (boulders, etc.) there are fundamental covers and the data itself is very incomplete. At the bottom right there is a situation after registering all the data for a given territory, where the situation is already considerably better, however, the terrain coverage is definitely not entirely compact.

Drone images were processed in Agisoft PhotoScan ver. 1.2.5, in total 1,389. The computa-tion quality has been set to "High quality" when the images are oriented even when generating a point cloud. A total of 7 computers and a server in a common network solution were used for the calculation. The orientation was made together for all the images, but the territory had to be divided into a total of 9 parts to generate the clouds to be merged. If the cloud comput-ing was solved in areas larger than eventually selected, there was a lack of RAM (all

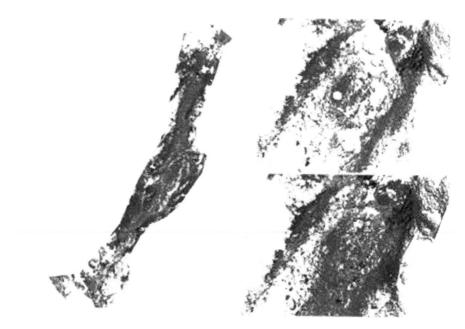

Figure 2.   3D scan data after registration - whole territory (left), top right detail - one measurement from the standpoint, bottom right coverage of the same territory after registration of all standpoints.

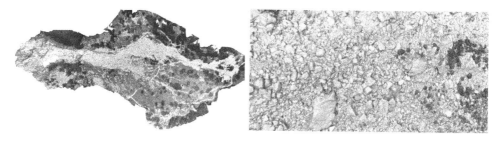

Figure 3.   Final data obtained by the SfM method (whole and detail).

computers with 64 GB RAM). The areas were chosen gradually with one unchanged boundary. A total of 261 million points were obtained after the data was merged.

The chosen solution procedure, where orientation was performed for all images and only the dense cloud generation was done in parts, was to ensure that the data is integral and there are no holes or overlaps. The achieved measurement quality was checked after the calculation, RMSE at the control points was less than 10 mm for all coordinates, and 0.16 pix on the image.

The total data obtained from the SfM method is shown in Figure 3 (left general overview, detail showing terrain capture detail right). Already the detail of the terrain capture is practically without holes, where the coverage of the space indicates a greater suitability of the method in such a specific terrain.

5   EVALUATION

Two point-clouds acquired by the different methods (3d scanning and UAV photogrammetry). Ground control points were determined with use of the same geodetic network.

The aim of the study is to compare these data and then analyze the accuracy of the methods in describing the very rugged terrain of the landslide. Basically, due to the direct nature of the polar 3D scanner measurement and the known measurement accuracy (the standard deviation of the spatial position determination is less than 3 mm, the measurement was less than 50 m) and the assuming that point accuracy of the control points was about 10 mm (at a maximum distance of 250 m), there can be assumed that the standard deviation of a single point location on the terrain will be of 10 mm at maximum. Thus, 3D scanning can be considered a more accurate method. A similar standard deviation can be used to characterize the position of the control points for the SfM method and thus the 3D scanning method can be considered more accurate.

Data comparison was done in CloudCompare ver. 2.10.2 [10], and due to the discontinuity of the 3D scan data, it was necessary to perform the calculation by selecting SfM data as a reference, and 3D scan data was projected on them. If done in reverse, photogrammetrically obtained data would show high deviations in the places of 3D scan data holes due to the absence of data. To calculate the distance of the surfaces and not just the individual points, the distance was determined relative to the local irregular triangular network formed between the 9 closest points. It would be possible to compare the point cloud directly with the triangular network, but the creation of a complex triangular network is very computational and memory demanding and the number of points would have to be greatly reduced, so this way was chosen. Before the comparison, the points representing vegetation (scrubs, grasses, and other lower fine growths) were removed from the clouds. The absolute distances, as well as the individual components of this distance in the direction of the X, Y, and especially Z coordinate axes were calculated. The height component is very important for the resulting data quality assessment, respectively. their mutual consent. The "Compute cloud-to-cloud distance" function was used, the components of length in the X, Y, Z direction were calculated, where local surface modeling using a triangular mesh made of the nearest nine points was used. Graphically is the resulting comparison for the absolute distance shown in Figure 4 – left, on the right in detail.

The mean absolute distance of the clouds is 0.04 m and the standard deviation of the remaining deviations is 0.087 m. 8 quite large, on its right side it is obvious why. Height deviations on a typical monitored area surface are shown here. Higher deviations are clearly seen in the low vegetation region (dark color), and also dark deviations are shown in areas between stones where the methods being compared work differently, ie, SfM reconstructs using facets and 3D scanning directly measures and thus specific there may be deviations that are not due to inaccuracy as such but different calculation or measurement methods. After removing these points from the cloud based on the size of the deviation, there were 537 million points in the cloud, the average absolute deviation was 0.028 m and the standard deviation at an absolute

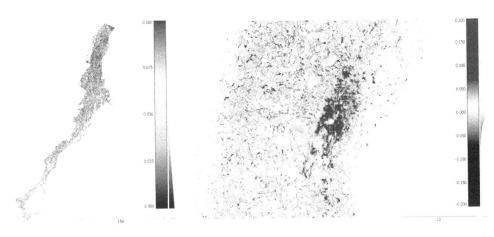

Figure 4.   The result of comparing the data of terrestrial 3D scanning and photogrammetric data (after manual removal of vegetation areas).

distance of 0.029 m. There was practically a negligible average shift of -0.008 m in the altitude component and a standard deviation of 0.032 m.

Aside from the difference in the acquisition of points of surface, the accuracy of the two methods compared is certainly sufficient to track changes in the observed area, at the level of centimeters. However, it is also necessary to assess the real applicability of the measurement methods, and here, despite the expected lower accuracy, imaging from drone and subsequent SfM photography is much more advantageous. In addition to easier time-consuming measurement, the advantage is also the complex coverage of the monitored area with virtually no holes, and the acquisition of very good color information to allow easy interpretation of any detected shifts.

6   CONCLUSIONS

The terrain is very specific, it is practically a stone field with boulders ranging from 3 meters in the longest dimension to stone sand. Furthermore, there is also a large slope of about 30% in the landslide direction. As a result of the comparison of the two methods used, the differences are characterized by a standard deviation of 0.03 m, indicating that, given the target of the tracking, both methods are certainly sufficiently accurate. Given the connection to the local system using geodetic measurements, it can certainly be assumed that the repeatability (i.e. comparability of data between stages) will also be at the same level.

In terms of laboriousness, there is a big difference between methods, drone imaging including stabilization of control points took about 3 hours. Excluding the aforementioned stabilization of control points, where it is not necessary to move in the monitored area, which is extremely advantageous due to its surface. In contrast, 3D terrestrial scanning took two business days, and the movement in the stone field with the measuring technique is very demanding and lengthy. Further, the low scanner position causes discontinuous terrain capture caused by many covers from nearby large boulders. When comparing the methods economically, drone imaging is also significantly cheaper, hardware and software costs are up to 10,000 Euros in the used configuration, while about 100,000 Euros in the case of 3D scanning (both without geodetic instruments for surveying geodetic surveying and control) points).

Furthermore, although the accuracy of both test methods is sufficient to track changes of this type, differences in measurement and data processing can cause differences in data that are not the product of terrain changes. For this reason, it is not advisable to compare the data obtained with these technologies if the aim is precision at the level of centimeters. Each of these technologies is in itself sufficiently precise, given the other parameters described above, it is possible to unequivocally recommend the most suitable drone imaging method, which is fast, accurate and environmentally friendly in this case.

REFERENCES

[1] Derrien, A. - Villeneuve, N. - Peltier, A. and Beauducel, F.: Retrieving 65 years of volcano summit deformation from multitemporal structure from motion: The case of Piton de la Fournaise (La Réunion Island), Geophys. Res. Lett., 2015, 42, 6959–6966, doi:10.1002/2015GL064820.

[2] Jovančević, S. D. - Peranić, J. – Ružić, I. and Arbanas, Ž.: Analysis of a historical landslide in the Rječina River Valley, Croatia. Geoenvironmental Disasters (2016) 3:26. DOI 10.1186/s40677-016-0061-x

[3] Rossi, G. - Tanteri, L. - Tofani, V. et al.: Multitemporal UAV surveys for landslide mapping and characterization. Landslides (2018) 15: 1045. https://doi.org/10.1007/s10346-018-0978-0.

[4] Ridolfi, E. - Buffi, G. - Venturi, S. and Manciola, P.: Accuracy Analysis of a Dam Model from Drone Surveys. Sensors 2017, 17, 1777.

[5] Buffi, G. - Manciola, P. - Grassi, S. – Barberini, M. and Gambi, A: Survey of the Ridracoli Dam: UAV–based photogrammetry and traditional topographic techniques in the inspection of vertical structures. Geomatics, Natural Hazards and Risk, 2017, 8:2,1562–1579, DOI: 10.1080/19475705.2017.1362039.

[6] Duró, G. - Crosato, A. - Kleinhans, M. G. and Uijttewaal, W. S. J.: Bank erosion processes measured with UAV-SfM along complex banklines of a straight mid-sized river reach. Earth Surf. Dynam., 6, 933–953, 2018 https://doi.org/10.5194/esurf-6-933-2018.

[7] Peppa, M. V. - Mills, J. P. - Moore, P. - Miller, P. E. and Chambers, J. E.: Brief communication: Landslide motion from cross correlation of UAV-derived morphological attributes. Nat. Hazards Earth Syst. Sci., 17, 2143–2150, 2017. https://doi.org/10.5194/nhess-17-2143-2017.

[8] Salvini, R. - Mastrorocco, G. - Esposito, G. - Di Bartolo, S. - Coggan, J. and Vanneschi, C.: Use of a remotely piloted aircraft system for hazard assessment in a rocky mining area (Lucca, Italy). Nat. Hazards Earth Syst. Sci., 18, 287–302, 2018. https://doi.org/10.5194/nhess-18-287-2018

[9] Kovacevic, M. S. - Car, M. - Bacic, M. - Stipanovic, I. - Gavin, K. - Noren-Cosgriff, C. and Kaynia, A.: Report on the use of remote monitoring for slope stability assessments. H2020-MG 2014-2015 Innovations and Networks Executive Agency.

[10] http://cloudcompare.org/, 2019.

Advances and Trends in Geodesy, Cartography and Geoinformatics II –
Molčíková, Hurčíková, & Blišťan (eds)
© 2020 Taylor & Francis Group, London, ISBN 978-0-367-34651-5

Determination of variation components of measurement from adjustment results

M. Štroner & O. Michal

Department of Special Geodesy, Faculty of Civil Engineering, CTU in Prague, Czech Republic

ABSTRACT: The geodetic network adjustment results can be used to determine the accuracy of the used measurements, but in case where nonhomogeneous measurements (dimensions, accuracy) are used, the estimation of standard deviations characterizing the variation of each type of measurements is not unambiguous. The reason, why to use this type of precision estimation is clear, the estimate of such determined accuracy should be much closer to reality than determining the accuracy of a single quantity separately (e.g. length baseline measurement, horizontal direction baseline measurement), including all types of errors including systematic ones. The problem is the determination of individual variation components, i.e. accuracy of individual measured variables, because their influence is mixed in the processing of the measurement by the least squares method and it cannot be easily separated. At the same time, the adjustment result is already pre-influenced by the used a priori standard deviations. Some theories and methods for solving this problem have been discussed in the literature, but they are very complicated, and some of them even fail in some cases. In the paper, the effect of variational component changes on a posterior unit standard deviation is presented for the case of terrestrial two dimensional (X, Y) geodetic network. After clarification of the dependence and some other important properties, a new method of variation components estimation using simple numerical approach is presented. Efficiency of the method is presented on modelled data evaluation examples and also on geodetic data measured in real.

1 INTRODUCTION

Accuracy of measurement of individual geometrical quantities is one of the basic parameters determining the usability of geodetic instruments. Reliable methods are known for determining the accuracy of one parameter under laboratory conditions, usually based on the use of a more accurate method. However, in practical use, such determined accuracy may not correspond to real values. Therefore, it is important to determine the accuracy of the measurements already made, ideally under the conditions corresponding to their actual processing, i.e. by the least squares method (LSM) adjustment. Measurement errors are influenced not only by the instrument, but also by external conditions (e.g. instrument stabilization and signaling, refraction, etc.), so the accuracy thus determined is more corresponding with reality and can be used, for example, for planning accuracy in similar cases in the future. Determining the accuracy of individual variables (or even their groups) is not, in principle, a simple mathematical task. The paper presents a simple numerical method of determining variation components (accuracy of individual groups of measured variables, e.g. horizontal directions, distance), which is based on the decomposition of aposterior unit standard deviation into components according to individual variation components. Compared to the existing known methods, no simplification is used and, in principle, the calculation cannot fail.

2 STATE OF THE ART

Many authors have dealt with this issue, eg [1–7]. In particular, older authors have attempted to derive universal finite methods, based on various complicated linear simplifications, but fail in the case of more complicated network configurations, such as negative variances. Here, as in all statistical and adjustment methods, the more unnecessary measurements, the better (more accurate) are the results, and thus the limited methods have a very limited applicability. For these reasons, we have tried to look at the problem from the point of view of the fundamental importance of the individual characteristics of accuracy and a method has been proposed respecting only these principles.

3 A NEW METHOD PROPOSAL

The basic indicator of the correctness of the estimated accuracy of individual measurements in the LMS is the aposteriori unit standard deviation s_0, which is determined from the formula:

$$s_0 = \sqrt{\frac{\sum pvv}{n-k}} = \sqrt{\frac{\sum pvv}{r}} \tag{1}$$

Here p is weight, v is correction, n is a number of all measurements, k is a number of necessary measurements, r is number of unnecessary measurements. The weight of the measurement is calculated from the a priori standard unit deviation (the chosen constant) σ_0 and the standard deviation of the individual measurement σ_i according to the formula:

$$p_i = \frac{\sigma_0^2}{\sigma_i^2} \tag{2}$$

If the weights are selected correctly, i.e. the standard deviations σ_i corresponds to the specified corrections v_i, then $s_0 = \sigma_0$ applies. For simplicity, $\sigma_0 = 1$ will be used. Therefore, the goal of a correct estimate is to have a state where, after alignment, $s_0 = 1$. If only one quantity is measured (eg only horizontal distances) and the same accuracy is attributed to all measurements, this is a trivial task. It is sufficient to calculate the alignment with weights equal to one and the resulting s_0 represents the accuracy of one measurement. In the case where different variables with different accuracy are measured (typically in the geodetic network horizontal directions, zenith angles and slope lengths), this procedure cannot be used. A certain estimate can be achieved by adjusting with an initial estimate of accuracy (e.g., according to the manufacturer's data), and after adjustment multiplying these standard deviations by s_0. This accuracy estimate is more credible, but fundamentally influenced by the initial estimate ratio that cannot change. The situation can be illustrated in Figure 1, where s_0 is shown for different variations of the horizontal direction and distance measurement standard deviations. For different tested networks graphs up to scale and shift looks the same. The correct solution can be any isocurve point $s_0 = 1$.

It is therefore necessary to consider that not only all measurements must have $s_0 = 1$, but this condition must in principle apply to all subgroups, i.e. e.g. in a planar geodetic network for horizontal directions $s_{0\varphi} = 1$ and for horizontal distances $s_{0d} = 1$. The essence of the solution is therefore the introduction of new accuracy characteristics - "a posterior unit standard deviation of the measurement group", which is calculated only from the corrections of the corresponding measurements, divided by the contribution of these measurements to the net overcapacity (their proportion of the unnecessary measurements), which is calculated using the reliability matrix R:

$$R = I - A\left(A^T P A + DD^T\right)^{-1} A^T P \tag{3}$$

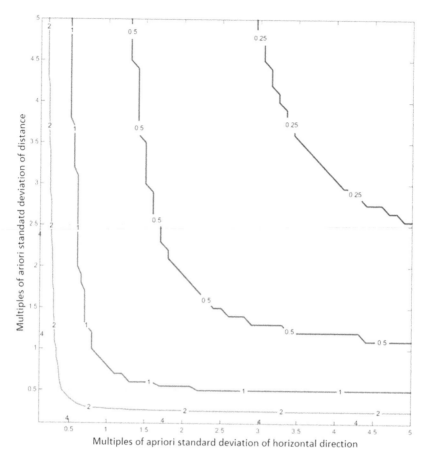

Figure 1. Isolines of aposteriori standard deviation s_0 depending on the used standard deviations of the measurements.

Here I is a unit matrix, A matrix of derivations, P matrix of weights, D matrix of geodetic network location. R is a square matrix with a dimension corresponding to the number of measurements. The trace of the matrix R corresponds to the number of redundant measurements $tr(R) = r$. It is therefore possible to calculate the number of redundant measurements for any group of measurements and thus also the aposteriori standard deviation:

$$s_0^{1sk} = \sqrt{\frac{\sum (pvv)_{1sk}}{r_{1sk}}} \tag{4}$$

Therefore, if the partial s_0 are determined numerically for the individual measurement groups in the adjustment, their common intersection of the isocurve (iso - surface, iso - hypersurface) $s_0 = s_{01sk} = s_{02sk} = \ldots = s_{0Nsk} = 1$ is the searched solution. The spatial graph of Figure 2 depicts changes in partial unit standard deviations (gray and black areas) and total s_0 (white checkered area) depending on changes in the size of the a priori prior aberrations. For the sake of clarity, only a portion of the most relevant results are shown, where total s_0 is between 0.8 and 1.2. In the case of the two standard deviations sought, these aposteriori values form surfaces that intersect with each other, and this intersection is also common to the overall a posteriori deviation.

94

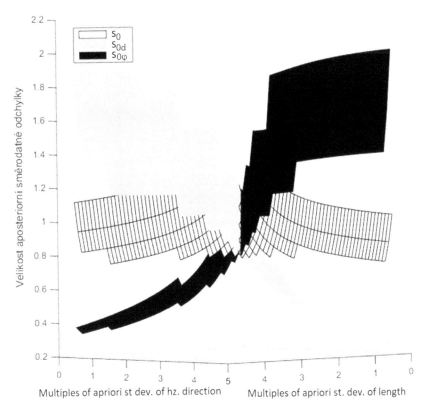

Figure 2. Graph showing the correlation of calculated standard deviations and their common intersection.

As a result, the estimation of variation components is at this point of intersection, at the point where all calculated aposteriori variation are equal to the a priori deviation. On this intersection, a point is found in which all the standard deviations are closest to one. The algorithm of numerical calculation can be summarized in a few points:

1. A priori standard deviation is selected for each measurement group (e.g., according to the manufacturer's specifications).
2. In the interval $1/x$ to x multiples of the a priori measurement deviations, the partial and the total deviations s_0 of the individual measurement groups are calculated (with 0.1-fold).
3. In the results, a combination of multiples is searched for, where all s_0 approach to 1 (increasing the interval from exact 1 to at least one combination is found, the average being used in the case of multiple combinations).
4. The procedure is repeated with the results obtained until the results change. (In case of networks tested by us, one repetition was always sufficient).

With a higher number of determined standard deviations, the procedure does not change basically, only a graphical representation would be more complicated, for example in the 3D geodetic network and measured values the slope distance, horizontal direction and zenith angle will not the aposteriori deviations $s_0 = 1$ form a surface, but a body. The common intersection of all the obtained bodies is the desired combination of variation components.

4 TESTING OF THE ALGORITHM

The proposed algorithm was practically tested, in order to control the results obtained, not a real measurements were used, but measurements with a pseudo-random error (corresponding to normal distribution) were generated. The geodetic network shown in Figure 3 was used, where points 11, 12, 13 are free standpoints, points 1 and 2 are fixed and points 3 to 10 determined. The network contains a total of 13 points, 3 orientation shifts, 64 measurements, 29 unknown unknowns, 35 redundant measurements.

4.1 Test 1 – repeated generation and calculation of variation components

As a first test, a reliability test was performed by repeatedly generating and calculating variation components, and the results are summarized in Table 1. The results prove the applicability of the method. When interpreting the results, it must also be considered that the generation of the pseudo-random normal distribution is not perfect and thus the set of measurements does not have to have the normal distribution with the exact standard deviation used to generate. The confidence interval was determined based on χ^2 probability distribution according to [8] with probability 95%.

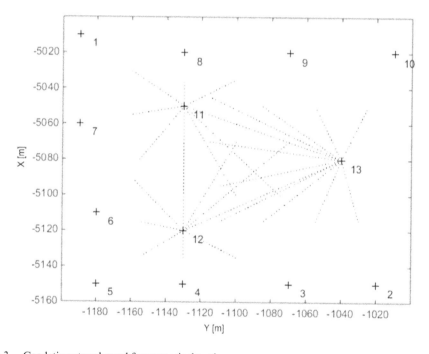

Figure 3.   Geodetic network used for numerical testing.

Table 1.   Results of the test 1.

Parameter	Generated precision	Confidence interval	Number of matching results	Average estimate	Maximal deviation
Horizontal distance [mm]	4,5	3,5 – 6,3	96%	4,4	–1,5; +0,6
Horizontal direction [mgon]	3,0	2,2 – 4,4	100%	3,1	–0,4; +0,6

4.2 Test 2 – changing the precision

In above described network the measurements were always generated with a particular combination of standard deviations that gradually shifted linearly in ten steps, standard deviation of direction 6 mgon to 0.6 mgon; standard deviations of 0.9 mm to 9 mm in length. Not all possible combinations were used, but only combinations matching the order. For each combination, a 10x generation and calculation of the variation components was always performed, the results are shown in Figures 4 and 5.

In Figures 4 and 5, the crosses are marking the individual determinations of the variation components, indicating the variance of the calculation. The solid black line indicates the correct (apriori) values, the gray line shows the confidence interval by the χ^2 distribution with a 95% probability according to [8]. Due to the relatively small number of redundant measurements, the results are in line with expectations.

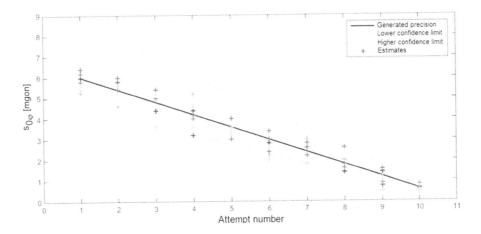

Figure 4.   Horizontal directional accuracy estimation results.

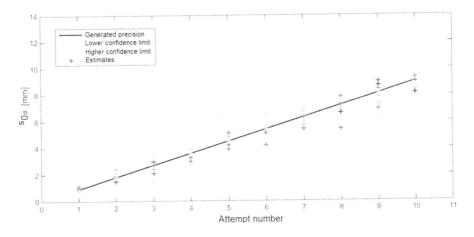

Figure 5.   Horizontal distance accuracy estimation results.

5 CONCLUSIONS

Accuracy is a very important characteristic of geodetic measurements, its determination is one of the basic tasks in accurate work. It is also worth noting that the result of any least squares adjustment is directly influenced by the weights used, which are directly determined by the standard deviations of the measurements, that is, the precision is a very important information for the correct use of the geodetic measurement results. The method of determining variation components from adjustment, which is based on numerical calculation and cannot inherently fail even for very large and complex networks with multiple measurements, has been shown. The functionality of the method has been demonstrated in several numerical tests; in the future, testing of the method on cases with multiple variation components is planned, for the beginning three (horizontal directions, zenith angles, slope distances). The only difference is that three types of measured values (horizontal directions, zenith angles and oblique lengths) are adjusted, so three accuracy characteristics will be determined

REFERENCES

[1] Kubáček, L.: Multivariate Statistical Models Revisited. Olomouc: Univerzita Palackého, 2008. ISBN 978-80-244-2212-1.
[2] Garafarend, E. W. – Schaffrin, B. (1979). Variance-covariance component estimation of Helmert Type. Surv. Mapping, XXXIX: 225–234.
[3] Horn S. D., Horn R. A., Duncan DB (1975). Estimating heteroscedastic variances in linear models. J. Am. Statist. Assoc., 70: 380–385.
[4] Persson C. G. (1981). On the estimation of variance components in linear models and related problems. Royal Inst. Technol. Stockholm.
[5] Rao, C. R. (1971). Estimation of variance and covariance components MINQUE theory. J. Multivariate Anal., 65: 161–172.
[6] Yavuz, E. – Baykal, O. – Ersoy, N.: Comparison of variance component estimation methods for horizontal control networks. International Journal of the Physical Sciences Vol. 6(6), pp. 1317–1324.
[7] P. J. G. Teunissen - A. R. Amiri-Simkooei: Least-squares variance component estimation. J Geod (2008) 82:65–82. DOI 10.1007/s00190-007-0157-x.
[8] Hampacher, M. - Štroner, M.: Zpracování a analýza měření v inženýrské geodézii. 2. vyd. Praha: Česká technika - nakladatelství ČVUT, ČVUT v Praze, 2015. 336 s. ISBN 978-80-01-05843-5.

Testing of drone DJI Phantom 4 RTK accuracy

R. Urban, T. Reindl & J. Brouček
Department of Special Geodesy, Faculty of Civil Engineering, CTU in Prague, Prague, Czech Republic

ABSTRACT: Technology of aerial photogrammetry using drones are widely used today and allow very detailed capture of virtually any objects, including the Earth's surface. The location and dimension of the resulting point cloud can be realized using ground control points (GCP) or by determining the spatial coordinates of the camera. The DJI Phantom 4 RTK is equipped with a GNSS receiver and its coordinates can be determined using the RTK method. In recent months, several studies have already been published on the accuracy of DJI Phantom 4 RTK, but mostly only give precision from multiple flights. The paper deals with testing the accuracy of the camera position during the air mission and also the accuracy of the points of the resulting point cloud, which is determined using RTK only. Testing was performed for different flight levels and for manual and automatic shooting modes. Phantom 4 RTK drone it can be reached results with an absolute error at a position of about 20 mm - 25 mm without any connection to the GCPs. Unfortunately, the evaluation of the absolute error for height is very problematic, as its value ranges from 8 mm to 129 mm.

1 INTRODUCTION

In these days the unmanned aircraft vehicle (UAV) are currently being used in many sectors of science (Fraštia, 2014), (Řezníček, 2013), (Blišťan, 2016), (Koska, 2017) and due to their economic aspects (Kršák, 2016), are increasingly available in the commercial sphere. Most of these systems use images from camera and basic principle of processing is epipolar geometry with image correlation (Štroner, 2013). For processing is typically used specialized software for aerial photogrammetry (Bartoš, 2014). It is possible to obtain a dimensionless or real spatial model from the image data. When creating a real spatial model, it is possible to use the ground control points (GCPs) on the surface of the scanned scene or directly the spatial position of the drone.The paper discusses the absolute accuracy of the DJI Phantom 4 RTK with the GNSS receiver, which allows the use of the RTK method. First, the accuracy of the drone spatial position, as indicated by the GNSS receiver, was assessed. Furthermore, the accuracy of the points on the calculated model determined from the positions of drone given by the GNSS receiver was also determined. Testing was done for different flight heights with both manual (the drone was stopped before the picture was taken for better quality of picture as well as for testing the GNSS antenna) and automatic shooting.

2 LOCATION

For testing purposes, a site located in the Czech Republic near the capital city of Prague was selected. The site is made up of arable land and the measurement took place only on its part with an approximate range of 1 - 1.5 ha. The area was not covered with vegetation, and perfect visibility was guaranteed for shooting GCPs during test of drone. 100 control points (Figure 1) were deployed in the selected area to form an approximate square about 100 m x 100 m (ie, the points were placed after about 10 meters). The GCPs were numbered from 1 to 100 and were

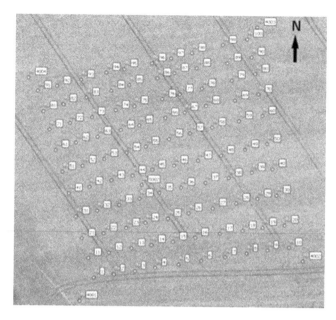

Figure 1. Placement of GCPs and points of geodetic network.

made using paper sheets of approximately A4 size with a central black and white divided 0.10 m radius, which provided optimal center detection for all flight types.

Four points of the geodetic network (4001-4004) were temporarily formed around the square grid (on diagonals), which were stabilized with wooden pegs. These points were used for later connection of GCPs to local Datum of Uniform Trigonometric Cadastral Network and height above sea level (Baltic Vertical Datum – After Adjustment). On the Figure 1 is also point 5002, which indicates the position of the total station determined by the free station from the points of geodetic network.

The position of the GCPs was determined by the spatial polar method at two positions of total station with the Trimble S6 HP (point accuracy of about 3 mm) from point 5002. Points 4001 - 4004 were re-determined before each drone flight. RTK method was realized with observation time of 15 seconds and connection to the CZEPOS virtual reference station.

3   DJI PHANTOM 4 RTK

Phantom 4 RTK is a small four-propeller drone from DJI with RTK module (Figure 2). On a single battery charge, the drone will last approximately 30 minutes in the air. It has a 20 megapixel camera with 1 inch CMOS sensor. The lens has a viewing angle of 84 ° with f/2.8 - f/1.1 focal length and a focus range of 1 m to infinity. The maximum video resolution is 3840x2160 pixels at 30 frames per second. It is equipped with a GNSS position module that allows reception of GPS (L1/L2), GLONASS (L1/L2) and Galileo (E1/E5a) satellites. The reported accuracy (for the GNSS position module) of the horizontal position determination is 1 cm 1 ppm and 1.5 cm 1 ppm. The accuracy of the autonomous flight of the drone when the RTK is turned on is in the position and at the height of 0.1 m. In addition to the classic buttons, the transmitter also has a display device on which the Android operating system is installed to operate the DJI MG application.

A total of 5 flights (Table 1) were carried out using a drone in three flight levels (40 m, 25 m and 15 m) and in two variants of shooting (manual and automatic). The transverse and longitudinal overlaps were set to 70 percent during automatic flight. In manual control, the orientation of the drone during the flight was fixed by operator and there was the same in all time.

Figure 2.   DJI Phantom 4 RTK.

Table 1.   The list of flight.

Flights	Number of photos [pcs]	Ground resolution [mm/pixel]	Reprojection error [pixel]
Automatic 40 m	213	12.3	0.45
Automatic 25 m	263	7.9	0.44
Manual 40 m	86	12.2	0.41
Manual 25 m	105	7.7	0.32
Manual 15 m	188	4.7	0.29

4   DATA PROCESSING

Processing has always been done similarly with the same outputs. Agisoft Photoscan v. 1.4.3 was used for photogrammetric outputs. The output coordinates from both the GNSS receiver and the drone were stored in WGS84 and subsequently transformed with the same key into local Datum of Uniform Trigonometric Cadastral Network and height above sea level (Baltic Vertical Datum – After Adjustment) in the Easytransform software.

4.1   *Accuracy of camera position*

First, the accuracy of the cameras for each flight is compared and examined. Accuracy of camera position for one flight specifies the spatial deviation of the image taken (camera position) relative to the GCPs. In the Agisoft software settings, the camera positioning accuracy was selected as very small and precise GCPs were used for the entire calculation. From the calculation it was then possible to export the spatial position of each camera, which was compared with the GNSS receiver position in the drone.

4.2   *Accuracy of point position on photogrammetric model*

Subsequently, the achieved accuracy of GCPs for individual flights was compared and examined. Accuracy of GCPs (in software) for one flight indicates with which spatial deviation the GCPs were calculated. In other words, as the Phantom 4 RTK drone could accurately determine imaginary points in the field without using GCPs. In this case, the camera positioning accuracy was set in Agisoft software as high-precision and the control points were kept as low-precision

5 RESULTS

For all the resulting images, flights at a height of 25 m above ground were chosen for clarity. The accuracy of individual models was influenced mainly by the determination of GCPs in the images, which is to some extent described as a reprojection error in Table 1. From these data, it is clear that the reprojection error was in fact at the level of a few millimeters with an increasing trend depending on the increase in flight level. Furthermore, the effect on accuracy can be attributed to the overlap of images, which is shown for 25m flight level in Figure 3 for an automatic variant and Figure 4 for manual. These data shows significantly better coverage in automatic flight, which is confirmed by the number of images in Table 1.

In contrast, it should be noted that in manual flight, the images are taken in a static drone position, while in an automatic drone they do not stop and images may be blurred. Last but not least, the overall photogrammetric model is somewhat plastic and in some cases the model from the higher flight level is better than from lower level. The Model from lower level may be deformed. Differences in individual coordinate axes were calculated for accuracy evaluation and then average values was determined (Xerror, Yerror, XY-error, Zerror and Total error). To eliminate outliers, the robust L1-norm method was used (Třasák, 2014).

5.1 Accuracy of camera position

The camera's position accuracy for 25 m is shown in Figure 5 for the automatic variant and Figure 6 for manual. The both figures shows the ellipses of errors for the XY component and for the Z component the ellipse is color coded according to the color scale always shown to the right of the image. The figures show an identical twist for a manual flight variant, indicating a systematic error probably in the reception of corrections (for RTK method) or in a different position of the GNSS receiver and the camera on the drone. To some extent, this error can also be caused by the same camera orientation that is not optimal for calculating the camera's internal orientation.

The parameters of all flights are shown in Table 2, where the positioning component (XY) achieves nearly values of precision determination by GNSS RTK method. The altitude component (Z) had better results from higher flights, but the accuracy is several times worse than the RTK method would have expected.

5.2 Accuracy of point position on photogrammetric model

The point position accuracy on the 25 m model is shown in Figure 7 for an automatic variant and Figure 8 for manual. The figure shows the ellipses of errors for the positioning component

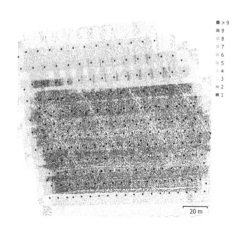

Figure 3.   Coverage - Automatic 25 m.

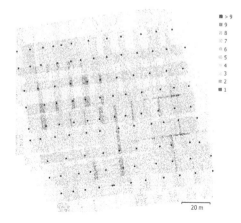

Figure 4.   Coverage - Manual 25 m.

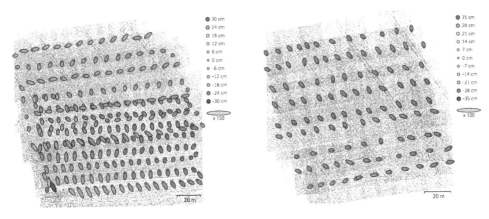

Figure 5. Cameras - Automatic 25 m.   Figure 6. Cameras - Manual 25 m.

Table 2. Average deviations in camera position.

Flights	X error [mm]	Y error [mm]	XY error [mm]	Z error [mm]	Total error [mm]
Automatic 40 m	15.9	37.9	41.1	122.4	129.2
Automatic 25 m	12.5	17.7	21.7	223.8	224.9
Manual 40 m	37.4	78.5	87.0	94.6	128.5
Manual 25 m	20.4	17.2	26.6	276.5	277.7
Manual 15 m	37.6	12.0	39.5	206.2	209.9

(XY) and for the height component (Z). The ellipses are again color-coded according to the color scale always shown to the right of the picture. Similarly as cameras, the images show the same twist for a manual flight variant. The parameters of all flights are shown in Table 3, where the positioning component (XY) achieves significantly better values for an automatic flight system where overlaps are maintained and there is a higher number of images than for manual. The height component (Z) in this case is between mm and hundreds of mm, which can only be explained by the RTK reception failure.

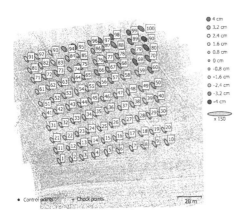

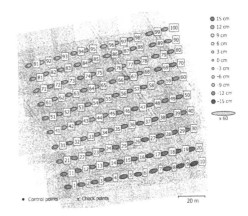

Figure 7. GCPs - Automatic 25 m.   Figure 8. GCPs - Manual 25 m.

Table 3. Average deviations in control point locations.

Flights	X error [mm]	Y error [mm]	XY error [mm]	Z error [mm]	Total error [mm]
Automatic 40 m	7.0	5.9	9.2	129.0	129.3
Automatic 25 m	12.9	6.5	9.9	78.0	78.6
Manual 40 m	50.2	70.2	86.3	8.3	86.7
Manual 25 m	45.6	10.2	46.7	109.9	119.4
Manual 15 m	12.1	11.4	22.5	22.5	27.9

6 CONCLUSION

The accuracy of Phantom 4 RTK from DJI was tested in the paper. The aim of the test was to find out and evaluate the absolute error for different flight levels in two flight variants. One variant of the flight consisted of an automatically controlled drone, where GNSS receiver had variable orientation. The second variant of the flight was manually controlled with the same GNSS receiver orientation. The data was evaluated in Agisoft PhotoScan.

First, the achieved accuracy of the cameras (for individual types of flights) was investigated, which determines the spatial deviation of the acquired images (camera positions) against the GCPs. In this testing, there was considerable discrepancy, especially in the height component (Z). This discrepancy can be caused by bad conversion of data from GNSS receiver to the camera position or bad corrections during the RTK method.

Then the accuracy of the GCPs (for individual types of flights) was tested, which determines how exactly a test drone can determine points without being connected to control points. It is therefore an information that is most interesting for practice. From these results, it can be seen that the positioning component (XY) achieves significantly better values for an automatic flight system where overlaps are maintained and there is a higher number of images compared to manual. The height component (Z) accuracy ranged from a few mm to hundreds of mm.

Overall, with the Phantom 4 RTK drone it can be reached results with an absolute error at a position of about 20 mm - 25 mm (for flight levels ranging from 15 m to 40 m) without any connection to the GCPs, assuming unequal orientation of the GNSS receiver during measurement. Unfortunately, the evaluation of the absolute error for height is very problematic, as its value ranges from 8 mm to 129 mm. It is therefore possible that there were problems with RTK corrections. To maintain greater accuracy, it is always preferable to make two flights with perpendicular paths.

ACKNOWLEDGEMENT

This work was supported by the Grant Agency of the Czech Technical University in Prague, grant No. SGS19/047/OHK1/1T/11.

REFERENCES

Štroner, M. & Pospíšil, J. & Koska, B. & Křemen, T. & Urban, R. & Smítka, V. & Třasák, P. *3D skenovací systémy*. Praha 2013: CTU Publishing House, p. 396, ISBN 978-80-01-05371-3.

Fraštia M. & Marčiš M. & Kopecký M. & Liščák P. & Žilka A. Complex geodetic and photogrammetric monitoring of the Kraľovany rock slide. In Journal of Sustainable Mining. Vol. 13, no. 4 (2014), s. 12–16. ISSN 2300-1364.

Bartoš K. & Pukanská K. & Sabová J. Overview of Available Open-Source Photogrammetric Software, its Use and Analysis. International Journal for Innovation Education and Research (IJIER). Vol. 2, no. 4 (2014), p. 62–70, ISSN 2201-6740.

Řezníček J. & Straková H. Documentation of Dumps and Heaps by use of UAV. Proceeding of 13th International Multidisciplinary Scientific GeoConferences SGEM 2, 151–158 (2013). ISBN 978-619-7105-01-8. ISSN 1314-2704.

Blišťan, P. & Kovanič, Ľ & Zeliznaková, V. & Palková, J. Using UAV photogrammetry to document rock outcrops. Acta Montanistica Slovaca, Volume 21 (2016), number 2, pp. 154–161, ISSN 1335-1788.

Kršák B. & Blišťan, P. & Pauliková, A. & Puškárová, P. & Kovanič, Ľ. ml. & Palková, J. & Zeliznaková, V. Use of low-cost UAV photogrammetry to analyze the accuracy of a digital elevation model in a case study. Measurement. Vol. 91 (2016), p. 276–287. ISSN 0263-2241.

Koska, B. & Jirka, V. & Urban, R. & Křemen, T. & Hesslerová, P. & Jon, J. & Pospíšil, J. & Fogl, M. Suitability, characteristics, and comparison of an airship UAV with lidar for middle size area mapping International Journal of Remote Sensing. 2017, 38(8-10), 2973–2990. ISSN 0143-1161.

Třasák, P. & Štroner, M. Outlier detection efficiency in the high precision geodetic network adjustment. Acta Geodaetica et Geophysica. 2014, 49(2), 161–175. ISSN 2213-5812.

Geodetic control and geodynamics

Advances and Trends in Geodesy, Cartography and Geoinformatics II –
Molčíková, Hurčíková, & Blišťan (eds)
© 2020 Taylor & Francis Group, London, ISBN 978-0-367-34651-5

Different view on the time series analysis of permanent GNSS stations

Ľ. Gerhátová & B. Hábel
Slovak University of Technology in Bratislava, Bratislava, Slovakia

ABSTRACT: The processing of data observed at the permanent GNSS stations results in 3D coordinates (X, Y, Z), including information on their uncertainty in the geocentric coordinate system with respect to its reference frame. They are presented typically in the form of daily or combined weekly solutions. For better geometrical representation and interpretation of the obtained 3D geocentric coordinates, we use topocentric coordinate changes in the North-, East- and Up-direction of the station position. The classical approach to time series analysis usually solves each of the components (n, e, u) separately as one-dimensional time series. The additive decomposition is the most commonly used. A stochastic noise process may be further analyzed. This paper introduces the possibilities of alternative time series analysis at the selected permanent GNSS stations. The horizontal position components reflect the trend movement of the Eurasian tectonic plate in the north-east direction of 27 mm/year. Based on the nature of these position components, we can analyze two-dimensional time series and model its common trend and a common seasonal or cyclical component. This approach was applied for the estimation of annual velocities at the selected permanent stations of the CEPER/ITMS network. The periodic components were studied using a spectral analysis method.

1 INTRODUCTION

The Global Navigation Satellite Systems (GNSS) and their observations provide, at various levels of accuracy, the ability to determine sites position on the Earth surface in real-time as well as in post-processing. In the case of continuous observations, we can obtain short-term and long-term information about time variability of point position. The actual accuracy of position is influenced besides the observations itself by many environmental factors affecting the geodetic monument stability. These are (Hefty 2004):

- Long-term position variations due to various global and plate tectonics in the form of annual velocity (several mm to cm per year).
- Disturbing environmental influences (periodic seasonal and short-term position variations) - the annual and daily variation of temperature, atmospheric pressure and humidity, direct solar radiation, wind strength and direction, underground water variability, and seasonal variations of vegetation surrounding the monument (level of a few mm).
- Short-term variations of position due to various geodynamic phenomena, e.g. tidal phenomena.
- Abrupt and irregular position variations associated with seismic activity.

GNSS observations (originally GPS only) from Slovakia have been available since 1991 to connect existing geodetic controls to the European Terrestrial Reference System 1989 (ETRS 89), and to determine the transformation parameters between ITRS/ETRS and geodetic coordinate systems in Czech and Slovak Republics. The oldest permanent station MOPI (Modra-Piesok) in Slovakia has begun to operate in May 1991. In 1996, it was included into the EPN (European Permanent Network). At the beginning of the 1990s, the project for the definition of new geodetic controls based on space and satellite techniques was created which resulted into the implementation of Slovak Positioning Service (SKPOS) with the permanent GNSS network. The MOP2 permanent station situated close to the existing MOPI was put into the operation in 2007 as a possible replacement for MOPI in order to provide a longer time series with aim determine the components of intra-plate velocities. In 2008, theMOP2 station was also incorporated into EPN.

The results of GNSS observations have the most often the form of time series which enable to investigate various geodynamic processes related to the location of the permanent station. In this study, we restrict our attention to the two-dimensional time series of topocentric coordinate changes of position in north (n) and east (e) directions which are the products of GNSS observations processing. We can examine each of these coordinates either separately as a one-dimensional series or together in the form of a two-dimensional data using the cointegration technique. This allows to adjust the motion of permanent stations including its direction.

2 COORDINATE TIME SERIES MODELING

A time series is a set of observations X_t recorded at a specific time t. The classical method used in time series modeling is a decomposition that deconstructs the original series into several components. In the case of additive decomposition, a time series model is in the form of (Brockwell & Davis 2002):

$$X_t = T_t + S_t + C_t + R_t, \qquad (1)$$

where T_t is a slowly changing function known as the trend, S_t is the seasonal component characterized by regular fluctuations based on the season with known and fixed period, and C_t is the cyclical component with periods other than a seasonal pattern. The residual component R_t has ideally the character of random noise and varies around zero.

2.1 *One-dimensional time series*

We assume that both coordinate time series n, e consist of linear trend and cyclical component involving all periodic fluctuations. The linear trend is the consequence of motion of the Eurasian tectonic plate at a rate of 27 mm/year. It may be expressed as a linear function of time independently for each coordinate series:

$$n_t^{trend} = a_n + b_n(t - t_0) \, \& \, e_t^{trend} = a_e + b_e(t - t_0). \qquad (2)$$

The coefficients a give the position of station at time t_0 and the slopes b of trend line may be interpreted as the rate of change in position as t changes (velocity). Similarly, we can approximate the cyclical component of time series composed of a sum of k individual periodic waves with known periods d_j as:

$$C_t = a_0 + \sum_{j=1}^{k} c_j \sin\left(\frac{2\pi}{d_j}t\right) + d_j \cos\left(\frac{2\pi}{d_j}t\right), \qquad (3)$$

where a_0 is the mean value of cyclical component with amplitudes:

$$A_j = \sqrt{c_j^2 + d_j^2}. \qquad (4)$$

In geodesy, the least squares method (LSM) is the most commonly applied for the adjustment of trend and cyclicity.

2.2 Two-dimensional time series

If the two-dimensional series n, e exhibit similar behavior, we can investigate their common trend and common cyclicity. This approach called a cointegration was originally described by Franses (1998). Some authors applied this methodology on the time series in geodesy and geodynamics, e. g Bognár (1999), Komorníková & Komorník (2002). The principle of cointegration is the use of transformation:

$$y_t = n_t \cos \alpha + e_t \sin \alpha \ \& \ x_t = -n_t \sin \alpha + e_t \cos \alpha. \tag{5}$$

Here y is a common trend direction and x corresponds to the trend-free series. Considering the topocentric coordinate system (n, e), the angle α represents the azimuth of common trend direction measured in the horizontal plane of ne. This angle can be derived from the ratio of deterministic trends which are defined in Equations 2 according to Komorníková & Komorník (2002):

$$\tan \alpha = b_e / b_n. \tag{6}$$

The transformed series y then allows to adjust the common trend from linear function:

$$y_t^{trend} = a_y + b_y(t - t_0). \tag{7}$$

Furthermore, the cyclicity in the residual series y after removing the common trend or in the trend-free series x may be approximated using the model described by Equation 3. The adjusted model of transformed series y, x involving the common trend and cyclicity is then:

$$y_t^m = y_t^{trend} + y_t^c \ \& \ x_t^m = x_t^c. \tag{8}$$

The model series y_m, x_m is possible to transform back to the (n, e) coordinate system using the backward transformation of cointegration.

3 EXPERIMENTS AND DISCUSSION

3.1 Experiment 1: Coordinate time series of MOP2 station

The permanent station MOP2 is situated in the area of the Astronomical Geophysical Observatory of the Comenius University (Bratislava, Slovakia) in Modra-Piesok in Little Carpathian Mountains. The monument of MOP2 was built in 2007 as a massive concrete pillar with the height of 3 m and diameter of 1 m. The monument stands on a concrete basement which is anchored to bedrock.

The weekly combined EPN position solutions in the SINEX format with their covariance information have been analyzed. They are based on the inputs from the EPN Analysis Centres (AC) downloadable from BKG GNSS Data Center (available at https://igs.bkg.bund.de/). Routine data processing of the EPN sub-network is tied to successive reference frames. The combination strategy for the generation of the combined EPN position solution is described in EPN (2019). Data processing until January 2017 was realized in the IGS08/IGb08 reference frames. The new realization ITRF2014 is used since January 2017 (EPN 2019). The analyzed data set gathered at the permanent GNSS station MOP2 covers a period of 8 years (from April 2011 to May 2019).

In the first part of the experiment we chose the classical approach to the time series analysis using the additive decomposition method. We considered each of the components of the horizontal position separately as a one-dimensional series (see Figure 1) and as the two-dimensional series applying the cointegration technique (see Figure 2). Adjusted trend parameters and amplitudes of cyclical components are listed in Table 1. For data analysis we used own scripts in MATLAB software. The significant frequencies of cyclical constituents were determined by the Lomb-Scargle periodogram due to the occurrence of gaps in analyzed series (Trauth 2006). The observed cyclicity is a combination of several fluctuations (a tropical and draconitic with period

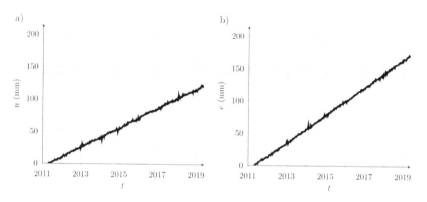

Figure 1. Topocentric coordinate time series of MOP2 station: a) north and b) east component.

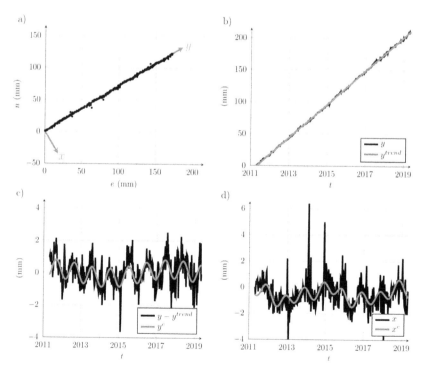

Figure 2. a) The north and east components of the MOP2 station in horizontal plane with transformation into the common trend direction y, b) time series of y fitted by liner trend y^{trend}, c) residuals of y after removing the linear trend and fitted by the cyclical component y^c, and d) trend-free time series x approximated by the cyclical component x^c.

Table 1.   Adjusted parameters of trend and cyclicity for the MOP2 station.

Component	Trend/σ (mm/y)		Period (d)	Cyclicity Amplitude/σ (mm)	
n	15.1	0.003	351.3	0.7	0.01
			1656.0	0.5	0.01
e	21.4	0.002	351.3	0.3	0.01
			1288.0	0.3	0.01
			644.0	0.2	0.01
y	26.2	0.003	351.3	0.6	0.01
			2318.4	0.3	0.01
x	0.0	0.000	362.3	0.5	0.01
			1449.0	0.4	0.01
$\alpha = 54.77041°$	$\sigma = 0.01463°$		644.0	0.2	0.01

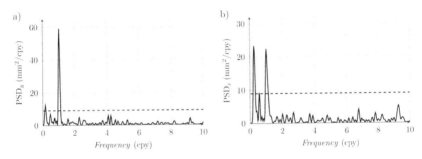

Figure 3.   Normalized power spectral densities (PSD) for: a) y and b) x transformed time series. Dashed line represents a power-level threshold corresponding to false-alarm probability of 5 % (cpy unit stands for cycles per year).

of 365.2 and 351.6 days, etc.), as we can see in Figure 3. This results from seasonal effects, data processing methods, uncertainty in a satellite positions, method of monumentation of the permanent stations, groundwater variability and some other local effects. The significance of the frequencies was assessed according to a power-level threshold corresponding to false-alarm probability of 5 %.

3.2   Experiment 2: Coordinate time series of CEPER/ITMS stations

The Central European Permanent Network (CEPER) currently includes 55 stations located in the Central Europe. Data processing in Bernese GNSS software (Dach et al. 2015) is similar to the EPN sub-network according to the recommendations and standards in guidelines for EPN Analysis Centres (EPN 2019). Since 2010, some other permanent stations have been gradually connected to the CEPER network from the project of National Center for Diagnosing the Earth's Surface Deformations in Slovakia (ITMS).

The cointegration approach is applied for selected permanent stations of the CEPER network and the ITMS project. Our aim was to determine the horizontal annual velocities of these stations and the azimuth of their movement which reflects the movement of the Eurasian tectonic plate. The list of stations is given in Table 2. The drift rates (velocities) are very similar at all tested stations and ranges from 25.5 to 27.2 mm/year. Anomalous behavior occurs at the GANP station (29.9 mm/year) due to the problem with antenna and receiver since January 2015. Uncertainty of trend estimates reflects the quality of time series (accuracy of

Table 2. Azimuths and annual velocities of the permanent station movements.

Station	Location	Azimuth/σ (°)		Trend/σ (mm/y)
BASV	Banská Štiavnica	56.32594	0.08960	26.4 0.04
BBYS	Banská Bystrica	56.28774	0.20013	26.7 0.09
BUTE	Budapešť (HU)	56.18826	0.08710	27.2 0.04
GANP	Gánovce	55.21267	0.06346	29.9 0.03
KAME	Kamenica and Cirochou	57.60330	0.09516	25.6 0.04
KOSE	Košice	55.95247	0.18026	26.9 0.08
LIE1	Liesek	55.43326	0.09166	25.8 0.04
MOP2	Modra-Piesok	55.57787	0.09499	26.4 0.04
PEMB	Partizánske	54.77811	0.09079	25.7 0.04
PENC	Penc (HU)	56.28842	0.08775	26.2 0.04
RISA	Rimavská Sobota	55.42959	0.18252	26.6 0.08
SUT1	Bratislava	51.85381	0.10335	25.5 0.04
TELG	Telgárt	56.34940	0.09996	26.6 0.04
TUBO	Brno (CZ)	54.54239	0.09019	25.6 0.04
USDL	Ustrzyki Dolne (PL)	54.00586	0.09081	25.8 0.04
ZYWI	Zywiec (PL)	55.20709	0.09297	25.7 0.04

individual weekly solutions, offsets or gaps in the data, etc.). The different azimuth estimate of the common trend was obtained at the SUT1 station. This permanent station is located on the roof of the building of the Faculty of Civil Engineering, Slovak University of Technology in Bratislava, Slovakia. Based on pillar instability monitoring experiment performed by precise inclination sensor (from March 2015 to March 2016), the SUT1 station shows daily changes in position depending on the season of the year. There is approximately twice higher variation during the summer than through the winter (Gerhátová et al. 2016.). Anomalous behavior due to the thermal accumulation in the summer and the tilt of the entire building is also reflected in the weekly combination of n component which has an impact on the estimate of the common trend azimuth and its accuracy. Different value of the common trend and azimuth estimated at the MOP2 station results from the use of different input data (a combined solution from several processing centers versus a separate solution).

4   CONCLUSIONS

In this paper we presented a different approach to the time series analysis, known as cointegration. It can be applied when two-dimensional data sets are available and the correlation between the individual components of the time series is strong. This is also the case of time series derived from the permanent GNSS observations, especially the topocentric coordinate changes of the station position in north and east direction. The main advantage of this approach is in its ability to determine the velocity of the permanent station movement and its direction in one step procedure as a common trend. In addition, the common cyclicity can be modeled. Practicability and usefulness of this approach was demonstrated on the network of selected GNSS stations where the estimated common trends were consistent with the movements of the Eurasian tectonic plate.

ACKNOWLEDGEMENTS

This work was supported by the Grants No. 1/0682/16 and 1/0750/18 of the Grant Agency of Slovak Republic VEGA. This paper is result of implementation of the project: National Centre for Diagnosing the Earth's Surface Deformations in Slovakia, ITMS 26220220108.

REFERENCES

Bognár, T. 1999. Time series analysis applied in geodesy and geodynamics. *J. Electr. Engrg.* 50(10): 19–23.

Brockwell, P.J. & Davis, R.A. 2002. *Introduction to time series and forecasting.* New York: Springer.

Dach, R., Lutz, S., Walser, P. & Fridez, P. 2015. *Bernese GNSS Software Version 5.2* [online], Bern: University of Bern, Astronomical Institute. Available: <http://www.bernese.unibe.ch>.

EPN – EUREF Permanent GNSS network 2019 [online]. Available: <http://epncb.oma.be>.

Franses, P.H. 1998. *Time series models for business and economic forecasting.* Cambridge: Cambridge Univ. Press.

Gerhátová, Ľ., Hefty, J. & Špánik, P. 2016. Short-term and long-term variability of antenna position due to thermal bending of pillar monument at permanent GNSS station. *Reports on Geodesy and Geoinformatics* 100(1): 67–77.

Hefty, J. 2004. *Global Positioning System in Four-dimensional Geodesy*, Bratislava: Slovak University of Technology.

Komorníková, M. & Komorník, J. 2002. Time series models for Earth's crust kinematics. *Kybernetika* 30(3): 383–387.

Trauth, M.H. 2006. *Matlab Recipes for Earth Sciences.* New York: Springer, 2006.

Comparison of different GRACE monthly gravity field solutions

J. Janák

Department of Theoretical Geodesy, Faculty of Civil Engineering, Slovak University of Technology in Bratislava, Slovakia

ABSTRACT: Satellite mission GRACE (Gravity Recovery and Climate Experiment) produced, during its 15 years of life time (2002-2017), unique measurements enabling to monitor the time variations of global gravity field of the Earth, typically with the time resolution of 1 month. The majomr part of the gravity variation is caused by hydrologic mass variation. Research presented in this paper is dedicated to comparison and statistical assessment of different GRACE monthly gravity field solutions available via International Centre for Global Earth Models (ICGEM) service. The Releases 5 and 6 from 3 official processing centres: University of Texas Centre for Space Research, German Research Centre for Geosciences, Jet Propulsion Laboratory are analysed. Paper also compares different types of filtration used in particular sets of the monthly gravity models. For spherical harmonic synthesis, the GrafLab software is used in Matlab computational environment. Mutual differences between the selected GRACE monthly gravity models are visualized and compared

1 INTRODUCTION

1.1 *Present status*

The new release RL06 of monthly gravity field models from all 3 official processing centers, German Research Centre for Geosciences (GFZ), see (Dahle et al. 2019), University of Texas Centre for Space Research (CSR) and Jet Propulsion Laboratory (JPL), has been issued in 2018. The new release should supplement or substitute the older release RL05 which has been used since 2016. The RL06 has been prepared in 2 different variants: variant BA containing the spherical harmonic coefficients up to degree and order (d/o) 60 and variant BB with the coefficients up to d/o 96. Table 1 comprises important basic information about the RL05 and RL06.

One should not forget that the basic parameters, geocentric gravitation constant GM and the length parameter R, of the GRACE monthly solutions are not always unified. In the case of official solutions presented in Table 1, the parameter $GM=0.3986004415 \cdot 10^{15}$ m³s⁻² is the same for all solutions, however, the parameter R is different for the GFZ solutions ($R=6378136.46$ m) and different for the CSR and JPL solutions ($R=6378136.30$ m). This heterogeneity causes slight systematic difference between the GFZ and CSR/JPL solutions depending on latitude. Although the difference is small, it is not always negligible as the time varying signal, which we are looking for, is also very small.

There have been several papers dealing with the comparison of particular monthly solutions either in a global scale (Sakumura et al., 2014) or using regional case studies (Jing et al., 2019), (Godah et al., 2016), (Klees et al., 2008). Some of these studies are focused primarily on analysis of different filtration methods, while other are concentrated on different processing strategies or additional sources of differences or try to combine the different solutions taking advantage of the particular solutions and thus minimize the systematic errors (Jean et al., 2018).

Table 1. Basic information about the GRACE official monthly solutions of RL05 and RL06.

Centre	Number of sets	Max. d/o	Number of files	Number of sets	Max. d/o	Number of files
		RL05 (2016)			RL06 (2018)	
GFZ	9	90/90	152	9 BA	60/60	133
				9 BB	96/96	133
CSR	9	90/90	149	9 BA	60/60	156
				9 BB	96/96	156
JPL	9	90/60	152	9 BA	60/60	156
				9 BB	96/96	147

1.2 Outline of the paper

Section 2 discusses the question of the choice of the reference model and quantifies the differences between particular reference models. Section 3 deals with the comparison of particular GRACE monthly solutions in South Asia region focusing on differences between the RL05 and RL06. Principal results are comprised in Conclusion.

2 CHOICE OF THE REFERENCE MODEL

2.1 What is the reference model

The reference model for studying the time varying gravity field should be stable in time and reasonably close to average of time varying models. If these conditions are met, the differences between the particular monthly models and the reference model clearly show time-varying part of the signal which can further be processed, analyzed and interpreted. However, because the time-varying signal is very small, relative to the whole gravity signal, even the small deviations of reference model from the above mentioned conditions can negatively affect the examining differences.

According to our investigation, it is not convenient to use the same reference model for all GRACE monthly solutions, because of different processing strategies used by particular centres for different releases and also due to different filtration parameters used for particular sets. All these factors may systematically influence the solution. We decided to use the average model of all monthly models of particular set as the reference model for that set. This means that we have different reference model for every set of the monthly models.

2.2 Differences between the reference models

First we computed the simple average models for GFZ, CSR and JPL RL05 sets filtered using the DDK1 decorrelation filter (Kusche et al., 2009). These average models represent the reference models for the particular set of GRACE monthly solutions. Differences between the average GFZ and CSR models, see Figure 1, and average GFZ and JPL models, are clearly latitude dependent due to different length parameter R and different processing strategies used by the computation centres. Differences between the CSR and JPL solutions of RL05 are more than one order lower and not correlated with the latitude, see Figure 2. Reference models, computed as the average models taken over full time span for particular set, for different filtration methods marked as DDK1 – DDK8 also differ significantly, mainly in Greenland and Antarctic Peninsula. Figure 3 shows the differences between the GFZ RL05 DDK1 and DDK2 reference models. At the end we compared the differences between the reference models for RL05 and RL06. We found the meridian stripe-wise pattern as it can be seen in Figure 4 for GFZ solutions. Differences between the RL05 and RL06 for CSR and JPL solutions are graphically similar. All examples shown in this paper were computed up to d/o 60.

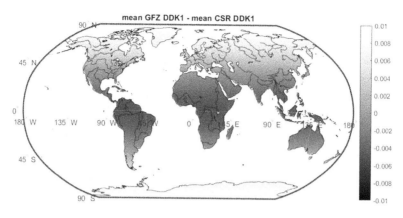

Figure 1.   Differences between the reference models GFZ and CSR of RL05 in terms of gravity. Differences between the GFZ and JPL of RL05 are graphically very similar. Units are 10^{-5} m.s^{-2}.

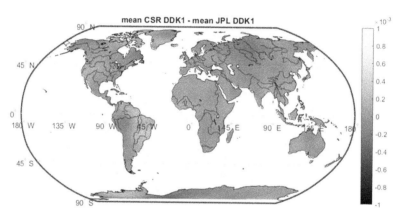

Figure 2.   Differences between the reference models CSR and JPL of RL05 in terms of gravity. Units are 10^{-5}m.s^{-2}.

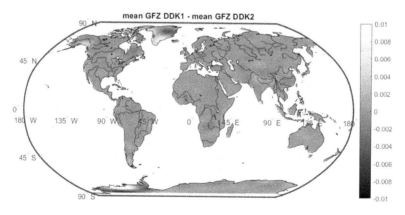

Figure 3.   Differences between the reference models GFZ RL05 DDK1 and DDK2 in terms of gravity. Units are 10^{-5}m.s^{-2}.

Table 1. Basic information about the GRACE official monthly solutions of RL05 and RL06.

	Number of sets	Max. d/o	Number of files	Number of sets	Max. d/o	Number of files
Centre	RL05 (2016)			RL06 (2018)		
GFZ	9	90/90	152	9 BA 9 BB	60/60 96/96	133 133
CSR	9	90/90	149	9 BA 9 BB	60/60 96/96	156 156
JPL	9	90/60	152	9 BA 9 BB	60/60 96/96	156 147

1.2 *Outline of the paper*

Section 2 discusses the question of the choice of the reference model and quantifies the differences between particular reference models. Section 3 deals with the comparison of particular GRACE monthly solutions in South Asia region focusing on differences between the RL05 and RL06. Principal results are comprised in Conclusion.

2 CHOICE OF THE REFERENCE MODEL

2.1 *What is the reference model*

The reference model for studying the time varying gravity field should be stable in time and reasonably close to average of time varying models. If these conditions are met, the differences between the particular monthly models and the reference model clearly show time-varying part of the signal which can further be processed, analyzed and interpreted. However, because the time-varying signal is very small, relative to the whole gravity signal, even the small deviations of reference model from the above mentioned conditions can negatively affect the examining differences.

According to our investigation, it is not convenient to use the same reference model for all GRACE monthly solutions, because of different processing strategies used by particular centres for different releases and also due to different filtration parameters used for particular sets. All these factors may systematically influence the solution. We decided to use the average model of all monthly models of particular set as the reference model for that set. This means that we have different reference model for every set of the monthly models.

2.2 *Differences between the reference models*

First we computed the simple average models for GFZ, CSR and JPL RL05 sets filtered using the DDK1 decorrelation filter (Kusche et al., 2009). These average models represent the reference models for the particular set of GRACE monthly solutions. Differences between the average GFZ and CSR models, see Figure 1, and average GFZ and JPL models, are clearly latitude dependent due to different length parameter R and different processing strategies used by the computation centres. Differences between the CSR and JPL solutions of RL05 are more than one order lower and not correlated with the latitude, see Figure 2. Reference models, computed as the average models taken over full time span for particular set, for different filtration methods marked as DDK1 – DDK8 also differ significantly, mainly in Greenland and Antarctic Peninsula. Figure 3 shows the differences between the GFZ RL05 DDK1 and DDK2 reference models. At the end we compared the differences between the reference models for RL05 and RL06. We found the meridian stripe-wise pattern as it can be seen in Figure 4 for GFZ solutions. Differences between the RL05 and RL06 for CSR and JPL solutions are graphically similar. All examples shown in this paper were computed up to d/o 60.

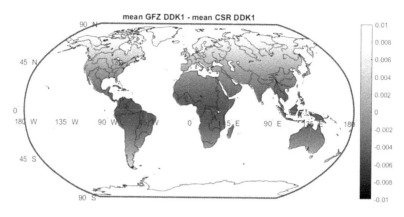

Figure 1.   Differences between the reference models GFZ and CSR of RL05 in terms of gravity. Differences between the GFZ and JPL of RL05 are graphically very similar. Units are 10^{-5} m.s^{-2}.

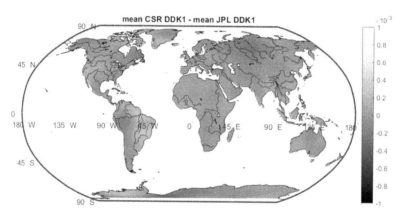

Figure 2.   Differences between the reference models CSR and JPL of RL05 in terms of gravity. Units are 10^{-5}m.s^{-2}.

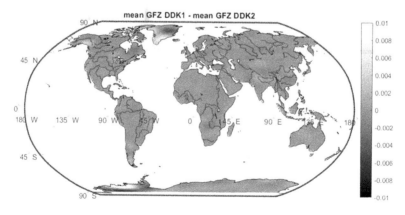

Figure 3.   Differences between the reference models GFZ RL05 DDK1 and DDK2 in terms of gravity. Units are 10^{-5}m.s^{-2}.

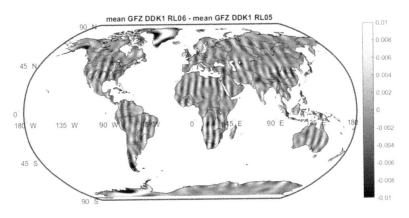

Figure 4. Differences between the reference models GFZ DDK1 RL06 and RL05 in terms of gravity. Units are 10^{-5}m.s^{-2}.

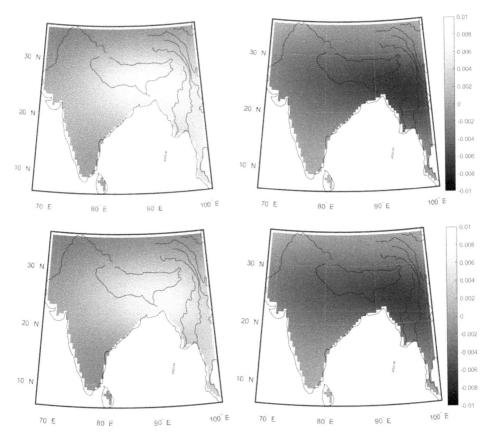

Figure 5. GFZ DDK1 solutions: for October 2004 RL05 (top left) and RL06, variant BA, (bottom left), for March 2010 RL05 (top right) and RL06, variant BA, (bottom right). Units are 10^{-5}m.s^{-2}.

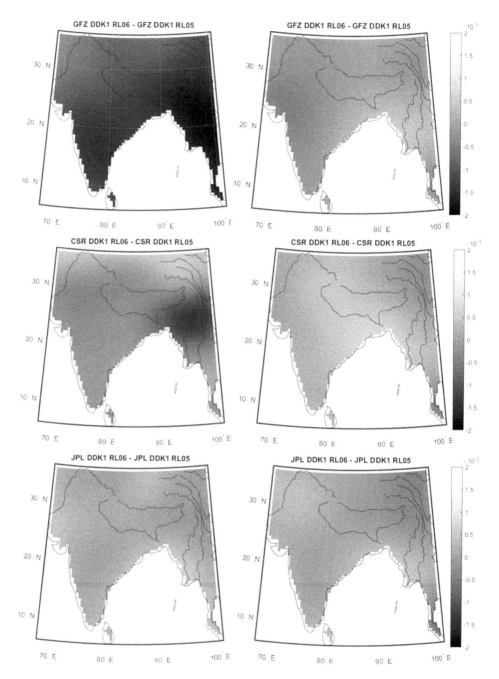

Figure 6. Differences between particular solutions of RL05 and RL06, variant BA: for October 2004 (left column) and for March 2010 (right column). Units are 10^{-5}m.s^{-2}.

3 CASE STUDY IN SOUTH ASIA

3.1 *Area description*

There are large and important aquifers in South Asia with huge depletion providing water for many millions of people and agriculture (Hirji et al., 2017), (Mukherjee et al., 2015), (Richey et al., 2015). We used the dry and wet periods in Ganges-Brahmaputra basin aquifer to make comparison of RL05 and RL06 and also different solutions of RL06.

3.2 *Numerical experiment*

The aim of the experiment was to compare particular monthly solutions tied to their own reference models focusing on the differences between the RL05 and RL06 in extremely wet and extremely dry months. We chose October 2004 as example of wet month, bright color in Figure 5, and March 2010 as example of dry month, dark color in Figure 5, for Ganges-Brahmaputra basin. Figure 5 presents the GFZ DDK1 solutions of RL05 and RL06 for selected months.

Their differences and also the corresponding differences between the CSR and JPL solutions are shown in Figure 6. We found out that the GFZ and CSR solutions of RL06 tends to have more moderate extremes than RL05 in South Asia region, which can be partially observed in Figure 6, dark colour of differences for October 2004 and bright colour for March 2010. We have also tested the differences between the particular sets (GFZ, CSR and JPL) of RL06 finding that these differences are smaller for selected region and selected extreme months than the differences presented in Figure 6.

CONCLUSION

In order to analyse relatively small monthly gravity changes it is convenient to use specific reference model for each particular set of GRACE monthly solutions. Sets prepared by CSR and JPL are in average closer to each other than to GFZ sets, see Figure 1 and Figure 2. Different parameters of decorrelation filters DDK1 – DDK8 affect the reference models mostly in polar regions, see Figure 3. Differences between the reference models of RL05 and RL06 show stripe-wise pattern in a North-South direction, see Figure 4. From the experiment presented in Section 3 we see that GFZ and CSR solutions of RL06 tends to have more moderate extremes than RL05 in South Asia region. Differences between the RL05 and RL06 are larger than differences between the GFZ, CSR and JPL solutions of RL06. This last finding is not demonstrated in the paper due to its limited extent.

ACKNOWLEDGEMENT

The research presented in this poster has been supported by the Slovak National Grant VEGA 1/0750/18 and by the Ministry of Education, Science and Sport of the Slovak Republic within the Research and Development Operational Programme for the project "University Science Park of STU Bratislava," ITMS 26240220084, co-founded by the European Regional Development Fund.

REFERENCES

Dahle, Ch., Murböck, M., Flechtner, F., Dobslaw, H., Michalak, G., Neumayer, K.H., Abrykosov, O., Reinhold, A., König, R., Sulzbach, R. & Förste, Ch. 2019. The GFZ GRACE RL06 monthly gravity field time series: Processing details and quality assessment. Remote Sensing 11:2116.
Godah, W., Szelachowska, M. & Krynski, J. 2015. On the selection of GRACE-based GGMs and a filtering method for estimating mass variations in the Earth system over Poland. *Geoinformation Issues* 7 (1): 5–14.
Hirji, R., Mandal, S., & Pangare, G. (Eds.). 2017. *South Asia Groundwater Forum: Regional Challenges and Opportunities for Building Drought and Climate Resilience for Farmers, Cities, and Villages*. New Delhi, India: Academic Foundation, 116pp.
Jean, Y., Meyer, U., & Jäggi, A. 2018. Combination of GRACE monthly gravity field solutions from different processing strategies. *Journal of Geodesy* 92 (11): 1313–1328.
Jing, W., Zhang, P., & Zhao, X. 2019. A comparison of different GRACE solutions in terrestrial water storage trend estimation over Tibetan Plateau. *Nature – Scientific Reports* 9:1765.

Klees, R., Liu, X., Wittwer, T., Gunter, B.C., Revtova, E.A., Tenzer, R., Ditmar, P., Winsemius, H.C., & Savenije, H.H.G. 2008. A comparison of global and regional GRACE models for land hydrology. *Surveys in Geophysics* 29: 335–359.

Kusche, J., Schmidt, R., Petrovic, S., & Rietbroek, R. 2009. Decorrelated GRACE time-variable gravity solutions by GFZ, and their validation using a hydrological model. *Journal of Geodesy* 83: 903–913.

Mukherjee, A., Saha, D., Harvey, Ch.F., Taylor, R.G., Ahmed, K.M., & Bhanja, S.N. 2015. Ground-water system of the Indian Sub-Continent. *Journal of Hydrology: Regional Studies* 4: 1–14.

Richey, A.S., Thomas, B.F., Lo, M.H., Reager, J.T., Famiglietti, J.S., Voss, K., Swenson, S., & Rodell, M. 2015. Quantifying renewable groundwater stress with GRACE. *Water Resources Research* 51: 5217–5238.

Sakumura, C., Bettadpur, S., & Bruinsma, S. 2014. Ensemble prediction and intercomparison analysis of GRACE time-variable gravity field models. Geophysical Research Letters 41: 1389–1397.

Advances and Trends in Geodesy, Cartography and Geoinformatics II –
Molčíková, Hurčíková, & Blišťan (eds)
© 2020 Taylor & Francis Group, London, ISBN 978-0-367-34651-5

Use of the topographic deflections of the vertical for computation of the quasigeoid

R. Kratochvíl, R. Machotka & M. Buday
Brno University of Technology, Brno, Czech Republic

ABSTRACT: The aim of this study is to open the possibility to substitute the gravimetric deflections of the vertical with the topographic ones in some cases. The topographic deflections of the vertical have a great advantage in the data availability for the calculation compared to the gravimetric ones. The detailed gravimetric data for the gravimetric deflections of the vertical calculation are not commonly accessible in many countries. On the contrary, topographic deflections of the vertical can be computed from the global digital elevation model with high resolution, for example SRTM (Shuttle Radar Topography Mission), or GMTED2010 (Global Multi-resolution Terrain Elevation Data 2010). These models are freely available.

The profile of Velká Bíteš – Brno – Uherské Hradiště in Czech Republic was chosen as the test area. There are mountainous as well as flat areas on the profile route. There is the greatest gradient of the height anomaly of the Czech Republic in this area. The profile consists of 581 points. On these points, three different types of deflections of the vertical were computed: gravimetric, topographic and deflections from the geopotential model EGM2008. The astrogeodetic deflections of the vertical were measured on 30 evenly distributed points of the profile and the GNSS/levelling was performed on 15 points of the profile. The terrain model GMDET2010 was chosen for the calculation of the topographic deflections of the vertical. The topographic deflections of the vertical were fitted on the deflections calculated from EGM2008. The fitted topographic deflections were compared with the gravimetric deflections, after that, the quasigeoid separation (height anomalies) on the profile was computed using the astronomical levelling method.

The root mean square error of the differences between the gravimetric and the fitted topographic deflections of the vertical is 0.53" in the meridian component ξ and 0.69" in the prime-vertical component η. The root mean square error of the differences in height anomaly calculated from the gravimetric and fitted topographic deflections of the vertical is 23 mm.

1   INTRODUCTION

This study deals with the possibility to substitute the gravimetric or astrogeodetic deflections of the vertical (DoV) with the topographic ones. Similar work was published by Hirt and his colleagues (Hirt et al. 2010), but they did not include the comparison with the gravimetric method. The topographic DoV can be computed from freely available global digital elevation models (DEM). On the other hand, the gravimetric data are not accessible in many countries. In this study the topographic DoV were computed from global DEM, the Global Multi-resolution Terrain Elevation Data 2010 (GMTED2010) (Danielson & Gesch 2011). The topographic DoV are relative values, therefore the DoV computed from Earth Gravitational Model 2008 (EGM2008) were used for their absolute positioning (Pavlis et al. 2008).

The EGM2008 model is complete to the degree and order 2160 with additional spherical harmonic coefficients up to the degree 2190 and order 2160. This means that the approximate spatial resolution of the model computed up to the full degree and order is 5'. This model is not able to capture shorter wavelengths of the gravity field. This missing part is often called an omission error (Gruber 2009). The high frequency waves of the gravity field, which are not possible to be captured by the global gravity models, are caused mostly by the topography of the Earth (Forsberg 1984). Two methods are available for the high frequency modelling of the gravity field. The first and more often used method is working with gravimetric data and Stokes or Molodenskij integral. The second one is based on the Newton integral applied to topographic masses modelled by DEM (Forsberg & Tscherning 1981).

The last mentioned method is the focus of this work. As a reference, other mentioned methods are also used. They serve as a source of independent data for evaluation of the precision of the results.

1.1 *The location*

The test was performed on the profile connecting the cities Velká Bíteš – Brno – Uherské Hradiště in the Czech Republic. The route of the profile was chosen purposely passing through territory with the biggest height anomaly gradient and on the same time passing through both mountainous and lowland areas. The profile begins in the west in the foothills of Českomoravská vrchovina, in the surroundings of Brno it passes through Dyjsko-svratecký úval and through Chřiby it continues to Dolnomoravský úval, where it ends. The sea level heights of the test profile are shown in Figure 1.

There are 30 astrogeodetic points in the profile. The astrogeodetic DoV on them were measured by the MAAS-1 system (Machotka 2013). The GNSS/levelling was performed on 15 selected points. Between each couple of the astrogeodetic points, 19 intermediate points were inserted. The geodetic coordinates of the intermediate points were determined by linear interpolation and their heights were obtained from the GMTED2010 model. Thus 581 points of the profile was created. At all these points the gravimetric as well as topographic DoV were computed.

The profile is 100 km long, the highest point is approximately 530 m above sea level, the lowest one 175 m above sea level.

2 DEFLECTIONS OF THE VERTICAL

Several different DoV sets were used as a source data for later computations and testing. They were: EGM2008 based DoV, topographic DoV, astrogeodetic DoV and gravimetric DoV. EGM2008 based DoV were computed using formulas of Torge and Müller (Torge & Müller 2012). They represent the low-frequency part of the DoV.

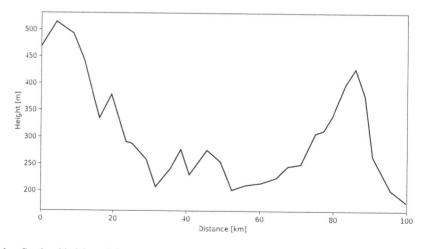

Figure 1. Sea-level heights of the test profile.

$$\xi = -\frac{GM}{r^2\gamma} \sum_{n=n_{min}}^{n_{max}} \left(\frac{R}{r}\right)^n \sum_{m=0}^{n} (\varDelta\bar{C}_{m,n} \cos m\lambda + \varDelta\bar{S}_{m,n} \sin m\lambda) \frac{\delta\bar{P}_{m,n}(\sin\varphi)}{\delta\varphi} \qquad (1)$$

$$\eta = -\frac{GM}{r^2\gamma\cos\varphi} \sum_{n=n_{min}}^{n_{max}} \left(\frac{R}{r}\right)^n \sum_{m=0}^{n} (\varDelta\bar{S}_{m,n} \cos m\lambda - \varDelta\bar{C}_{m,n} \sin m\lambda) m\bar{P}_{m,n}(\sin\varphi) \qquad (2)$$

Here ξ is the meridian component, η is the prime-vertical component of DoV, GM is the geocentric gravitational constant, r is the spherical radius, γ is the normal gravity, R is the radius of the reference sphere, $\varDelta\bar{C}_{m,n}$ and $\varDelta\bar{S}_{m,n}$ are fully normalized spherical harmonic coefficients after removal of the normal field of degree m and order n, $\bar{P}_{m,n}(\sin\varphi)$ is the fully normalized associated Legendre function of degree m and the order n, φ and λ are geodetic latitude and longitude. These components were computed at all points of the test profile.

The topographic DoV (high-frequency part of DoV) $(\xi, \eta)_{topo}$, was computed from GMTED2010 model. Resolution of the model is 7.5 x 7.5 arc-seconds, its height RMS error is approximately 27 m (Danielson & Gesch 2011). Topographic DoVs were calculated using cuboids according to (Nagy et al. 2000, 2002).

In the first step, the horizontal derivatives of gravitational potential of topographic masses V_x^{topo} and V_y^{topo} were calculated.

$$V_x^{topo} = G\rho \left|\left|\left| y \ln(z+r) + z \ln(y+r) - x \tan^{-1}\frac{yz}{xr} \right| \begin{matrix} x_2 \\ x_1 \end{matrix} \right| \begin{matrix} y_2 \\ y_1 \end{matrix} \right| \begin{matrix} z_2 \\ z_1 \end{matrix} \qquad (3)$$

$$V_y^{topo} = G\rho \left|\left|\left| z \ln(x+r) + x \ln(z+r) - y \tan^{-1}\frac{zx}{yr} \right| \begin{matrix} x_2 \\ x_1 \end{matrix} \right| \begin{matrix} y_2 \\ y_1 \end{matrix} \right| \begin{matrix} z_2 \\ z_1 \end{matrix}, \qquad (4)$$

where G is the Newton gravitational constant, $\rho = 2.670$ kg.m^{-3} is the standard density of topographic masses, x_i, y_i, z_i are the coordinates of the prism boundaries related to the running point of the integration and $r(x, y, z) = \sqrt{x^2 + y^2 + z^2}$, where x, y, z are the coordinates of the point of computation.

Next, the individual components of the high-frequency topographic DoV were derived using the following equations

$$\xi_{topo} = \frac{V_y^{topo}}{\gamma} \qquad (5)$$

$$\eta_{topo} = \frac{V_x^{topo}}{\gamma}, \qquad (6)$$

where γ is the normal gravity on the surface of ellipsoid GRS80.

The integrating radius for the calculation of topographic DoV was chosen to be 10 km. The radius was felt to be sufficient as high-frequency signal only was needed. The same radius was used in the calculation of gravimetric DoV.

The astrogeodetic DoV were determined on 30 points of the profile using the system MAAS-1 and their accuracy is 0.30" in both components of the DoV (Machotka 2013).

The gravimetric DoV ξ_{grav}, η_{grav} were determined by the combination of the global gravity model EGM2008 and the numerical integration of residual ground gravity anomalies related to the Earth's surface using Vening Meinesz formulas and integration radius 5' (Kostelecký et al. 2011). The RMS error of ξ_{grav}, η_{grav} is approximately 0.30" in both components (Machotka 2013).

3 FITTED DOV CALCULATION

In the first step a residual model of DoV (ξ_{rez}, η_{rez}) was calculated as a difference of EGM2008 based on DoV and GMTED2010 in 5 arc-minutes grid base DoV, as shown in the equations (7) and (8). This step eliminated the residual topography effect from EGM2008.

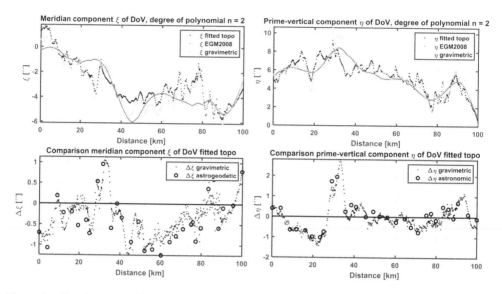

Figure 2. Fitted topographic, EGM2008 based, gravimetric and astrogeodetic DoV. The upper two graphs are showing the absolute values of DoV (fitted topo in the darkest, EGM2008 in the middle and gravimetric in the lightest shade). The reference gravimetric and EGM2008 based DoV are added for comparison. Lower two graphs are showing the differences between of fitted topographic DoV from gravimetric and astrogeodetic DoV.

$$\xi_{rez} = \xi_{EGM2008} - \xi_{GMTED2010}^{5'} \tag{7}$$

$$\eta_{rez} = \eta_{EGM2008} - \eta_{GMTED2010}^{5'} \tag{8}$$

The DoV residual model was fitted with a second-order polynomial to satisfy the least squares method condition. The result of this are the DoV compoments ξ_{pol} and η_{pol}. The final components of DoV ξ_{fitted} and η_{fitted} were computed as a sum of both ξ_{pol} and η_{pol} and a high-frequency of DoV computed from GMTED2010 in 7.5 arc-seconds grid, as shown in equations (9) and (10).

$$\xi_{fitted} = \xi_{pol} + \xi_{GMTED2010}^{7.5'} \tag{9}$$

$$\eta_{fitted} = \eta_{pol} + \eta_{GMTED2010}^{7.5'} \tag{10}$$

The resulting DoV are shown on Figure 2.

Good mutual agreement of fitted topographic and gravimetric DoV can be seen on Figure 2. Extreme differences and the RMS error of the comparison between the fitted topographic and gravimetric DoV as well as between the fitted topographic and astrogeodetic DoV are shown in

The comparison of fitted topographic with gravimetric and astrogeodetic DoV in Table 1. indicates that the RMS error of the method is approximately 0.70″. This value is a pessimistic estimate and does not take into account the RMS error of the reference data. The RMS error of the reference DoV is given in section 2. After applying of the error propagation law, the RMS error of determined fitted topographic DoV will be 0.60″.

4   QUASIGEOID CALCULATION

A astrotopographic quasigeoid was computed from the fitted topographic DoV by the astro-geodetic levelling according to (Molodenskij et al. 1962), extended by points with topographic DoV by (Hirt & Flury 2008). The Figure 3. shows comparison of astrotopographic quasigeoid

Table 1. The comparison of fitted topographic DoV with reference data.

| | Gravimetric vs. fitted topographic | | Astrogeodetic vs. fitted topographic | |
	$\Delta\xi$ [']	$\Delta\eta$ [']	$\Delta\xi$ [']	$\Delta\eta$ [']
Min	-1.24	-1.41	-1.25	-1.01
Max	1.13	2.80	0.96	1.95
Mean	-0.31	-0.04	-0.37	0.05
RMS	0.53	0.69	0.69	0.67

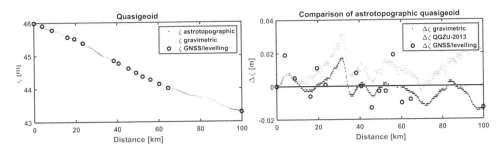

Figure 3. The comparison of astrotopographic quasigeoid with reference data. On the left, there is the progress in the route profile. On the right, there is the comparison of astrotopographic quasigeoid with gravimetric quasigeoid, quasigeoid QGZÚ-2013 and quasigeoid from GNSS/levelling.

Table 2. The comparison of astrotopographic quasigeoid with reference data.

	Gravimetric $\Delta\varsigma$ [m]	QGZÚ-2013 $\Delta\varsigma$ [m]	GNSS/levelling $\Delta\varsigma$ [m]
Min	-0.015	-0.004	-0.013
Max	0.017	0.032	0.019
Mean	-0.001	0.011	0.000
RMS	0.006	0.013	0.010

with reference data. GNSS/levelling quasigeoid separations (Kostelecký et al. 2004), gravimetric model based on local gravity data combined with EGM2008 (Kostelecký et al. 2011) and the Detailed gravimetric quasigeoid QGZÚ-2013 available on ČÚZK (ČÚZK 2013) were used for testing for the precision of the quasigeoid profile.

The comparison of astrotopographic quasigeoid with reference data is shown in Table 2.

The RMS error of the quasigeoid heights of the gravimetric quasigeoid indicates good agreement between the astrotopographic and gravimetric solutions of the profile. Given the RMS error of the quasigeoid heights QGZÚ-2013 over the ellipsoid, which is 1 cm (ČÚZK 2013) and the RMS of GNSS/leveling error which is 1.5 cm (Kostelecký et al. 2004), it can be concluded that accuracy of the astrotopographic quasigeoid is approximately the same.

5 CONCLUSIONS

The biggest deviation of the fitted topographic DoV and the astrotopographic quasigeoid from the reference model can be found around the 30th km of the profile, see Figure 3. This may be due to a sudden change in the density of the Earth's crust which is not taken into account in the calculation of topographical DoV. Here is the opportunity for calculation refining to reflect the density of the Earth's crust in vicinity of individual points. How well the described method would work for in the area network when the polynomial fit would be replaced by a quadratic surface remains an open question.

The astrotopographic levelling appears to be a good alternative to gravimetric methods in areas with a lack of gravimetric data, typically in mountainous or border areas. In these places, the authors of QGZÚ-2013 admit a larger model RMS error than in the central part of the model area. An appropriate combination of computational methods could make the model more homogenous.

The method provides relatively good results using commonly available data with the accuracy of 0.6″ for fitted topographic DoV and about 1 cm in quasigeoid heights for astrotopographic quasigeoid profile of 100 km length.

ACKNOWLEDGEMENT

This work was supported by FAST-J-15-2863.

REFERENCES

ČÚZK. (2013). "Podrobný kvazigeoid QGZÚ-2013." Retrieved 13.06., 2019, from https://geoportal. cuzk.cz/(S(inkw3xatgbduhirimblmdj55))/default.aspx?mode=TextMeta&side=bodpole&metada taID=CZ-CUZK-QGZU&menu=275, (in Czech).

Danielson, J. J. & Gesch, D. B. (2011), Global multi-resolution terrain elevation data 2010 (GMTED2010): U.S. Geological Survey Open-File Report 2011-1073, 26 p.

Forsberg, R. (1984), A study of terrain reductions, density anomalies and geophysical inversion methods in gravity field modelling: OHIO STATE UNIV COLUMBUS DEPT OF GEODETIC SCIENCE AND SURVEYING, 129 p.

Forsberg, R. & Tscherning, C. C. (1981). "The use of height data in gravity field approximation by collocation." Journal of Geophysical Research: Solid Earth 86(B9): 7843–7854.

Gruber, T. (2009), Evaluation of the EGM2008 gravity field by means of GPS-levelling and sea surface topography solutions: Institut für Astronomische und Physikalische Geodäsie, 15 p.

Hirt, C. et al. (2010). "Combining EGM2008 and SRTM/DTM2006.0 residual terrain model data to improve quasigeoid computations in mountainous areas devoid of gravity data." Journal of Geodesy 84(9): 557–567.

Hirt, C. & Flury, J. (2008). "Astronomical-topographic levelling using high-precision astrogeodetic vertical deflections and digital terrain model data." Journal of Geodesy 82(4-5): 231–248.

Kostelecký, J. et al. (2004). "Quasigeoids for the territory of the Czech Republic." Studia Geophysica Et Geodaetica 48(3): 503–518.

Kostelecký, J. et al. (2011), Technická zpráva: Výpočet gravimetrických tížnicových odchylek a výšek kvazigeoidu na bodech sítě AGNES: VÚGTK Zdiby, 4 p., (in Czech).

Machotka, R. (2013). Automatizace astronomického určování polohy Habilitation theses, Brno University of Technology, (in Czech).

Molodenskij, M. S. et al. (1962). Methods for study of the external gravitational field and figure of the earth, Jerusalem, Israel Program for Scientific Translations, 1962;[available from the Office of Technical Services, US Dept. of Commerce, Washington].

Nagy, D. et al. (2000). "The gravitational potential and its derivatives for the prism." Journal of Geodesy 74(7-8): 552–560.

Nagy, D. et al. (2002). "Erratum: corrections to: "The gravitational potential and its derivatives for the prism"." Journal of Geodesy 76(8): 475–475.

Pavlis, N. K. et al. (2008). An Earth gravitational model degree 2160: EGM2008. General Assembly of the European Geosciences Union. Vienna, Austria.

Torge, W. & Müller, J. (2012). Geodesy - 4th Edition. Göttimgen, Germany, Walter De Gruyter.

Advances and Trends in Geodesy, Cartography and Geoinformatics II –
Molčíková, Hurčíková, & Blišťan (eds)
© 2020 Taylor & Francis Group, London, ISBN 978-0-367-34651-5

The use of satellite radar interferometry to monitor landslides in Slovakia

M. Lesko & J. Papco
Department of Theoretical Geodesy, Faculty of Civil Engineering, Slovak University of Technology in Bratislava, Bratislava, Slovak Republic

M. Bakon
Department of Environmental Management, Faculty of Management, University of Presov in Presov, Presov, Slovak Republic

R. Czikhardt
Department of Theoretical Geodesy, Faculty of Civil Engineering, Slovak University of Technology in Bratislava, Bratislava, Slovak Republic

M. Plakinger
Mines of Upper Nitra Prievidza, a.s., Prievidza, Slovak Republic

M. Ondrejka
Department of Engineering Geology, State Geological Institute of Dionyz Stur in Bratislava, Bratislava, Slovak Republic

ABSTRACT: The paper deals with the use of satellite radar interferometry for monitoring landslides in Slovakia. Landslides represent the most risky type of natural danger in Slovakia, so there is a need to search for new monitoring options. In this case, satellite radar data was used to record the situation in the selected location in Slovakia. Specifically, it is the area of Upper Nitra, where landslides occur due to the geological structure of the area and subsidence due to mining activities. Two methods, Small Baseline Subset (SBAS) and Persistent Scaterrer Interferometric Synthetic Aperture Radar (PSInSAR), were used to find suitable data processing techniques, the processing results of which at the selected site in Slovakia are presented in this paper. The results of the radar satellite data processing so far point to the fact that it is necessary to consider the specific natural character of Slovakia's territory, which significantly distorts the resulting values of the movement of individual reflective elements. Therefore, the results in this paper are evaluated from the point of view of the unique relief, vegetation and atmospheric conditions of Slovakia, because they significantly influence the accuracy and quality of processing results.

1 INTRODUCTION

Satellite radar interferometry is a powerful remote sensing technique able to generate digital elevation models (DEMs) and measure displacements of the Earth´s surface. The second is a wide range of applicability: tectonic deformations, volcanic deformations, ground subsidence or lift caused by different causes (most often gas extraction or groundwater activity), landslides, glacial motion and monitoring stability of infrastructure and buildings.

Satellite radar interferometry methods, capable of generating time series, represent an extension of the classical InSAR (Interferometric Synthetic Aperture Radar) method and allow analysis of evolution or changes in the Earth's surface (Shanker et al., 2011). These multitemporal methods use extensive sets of satellite radar data to generate time series for individual scatterers. The advantage is that they are able to overcome some of the limits of the classical InSAR method. One such significant limit is the presence of uncorrelated phase noise as a result of the influence of the atmosphere on electromagnetic signal. The possibility of using more data results in the ability to partially reduce this effect, thus improving results. From a broader perspective, the mutlitemporal InSAR algorithms can be divided into two categories: Persistent Scatterer (PS) (Ferretti et al., 2001) and Small Baselline (SB) (Bernardino et al., 2002). The distribution is defined by the pixel selection method according to the reliability of the phase measurements (Shanker et al., 2011).

When applying the PS method, one image is a master image and all others are considered as slave images. Interferograms are generated between master image and individual slave images. The purpose of such generation of interferograms is to optimize the ratio between signal and noise within one pixel. A pixel includes a single dominant reflection element, the phase of which comprises disturbing additions of the reflected signal from all the reflection elements present in that pixel (Shanker et al., 2011). SB methods use interferograms generated between pairs of radar images obtained in a short time interval to maintain high temporal and spatial coherence. From the viewpoint of radar image distribution, there are three approaches to SB methods (Casu, 2009):

a) interferograms are generated between pairs of consecutive radar images (most radar images represent a slave image in one pair and a master image in another);
b) a first radar image is identified as master within the data set and the other is defined as slaves (similar to PS methods, but smaller data sets are used here to maintain high temporal coherence);
c) splitting the data file into smaller sets of radar frames in which the optimal combination of pairs is performed.

The SBAS algorithm utilizes distributed scatteres (DS) and singular value decomposition to connect independent multiple unwrapped interferograms in time (Bernardino et al., 2002), (Lanari et al., 2004). The master and slave pairs for these interferograms are selected using average baseline parameters for the signal of interest. In case of deformation analysis, baseline parameters can be set to 25% of the critical baseline (≈400 m) and ≈1 year for temporal baseline.

The short temporal baselines are especially needed for highly variable phenomena to minimize temporal decorrelation. SBAS modification, in which only interferograms with short temporal baselines are used, is called Small Temporal Baseline Subset (STBAS).

DS can be affected by baseline decorrelation, which happens more often in non-urban areas.

Atmospheric filtering is done by extracting the signal with high spatial and low temporal correlation using a spatio-temporal filter, as is the case for PSInSAR (Ferretti et al., 2000), (Ferretti et al., 2001).

Both types of methods (PS and SB) have been successfully applied in recent years to analyze various phenomena such as emerging or changing fault areas, landslides, changes in ground surface height due to changes in groundwater levels (Bürgmann et al., 2006), (Lanari et al., 2007). Independent studies demonstrating the accuracy of time series obtained by PS and SB approaches are described in (Casu et al., 2006), (Ferretti et al., 2007).

This paper compares two methods - the PSInSAR method (Persistent Scatterer Interferometric Synthetic Aperture Radar) based on the principles described in (Kampes, 2006) and the SBAS (Small Baseline Subset) method characterized in (Casu, 2009). Both methods were applied in the processing of satellite radar data by the Sentinel-1 satellite mission provided by the European Space Agency (ESA).

2 MONITORED AREA

The comparison of the results of radar data processing was carried out in the area of Upper Nitra (the area located in the northern part of the Nitra river basin) in the district of Prievidza.

The investigated localities were the villages of Hradec and Velka Lehotka (Figure 1), which were affected by slope failures between November 2012 and May 2013.

The geological structure is characterized by high sensitivity to anthropogenic interventions and at the same time it creates suitable conditions for rainwater infiltration. Landslides occurred on slopes formed by clays and clays containing deluvial sediments (Hradec) and alluvial sediments (Velka Lehotka).

The combination of anthropogenic activity and above-average rainfall following an unusually dry period led to the activation of slope failures. In connection with the fact that mining industry is being implemented in the southern part of the surveyed area, the site is subject to an engineering geological survey. In the framework of cooperation with the State Geological Institute of Dionýz Štúr and HBP Prievidza, Inc., the area represents a suitable example for the analysis of the possibilities of using satellite radar interferometry to monitor the movements of the Earth's surface in Slovakia. Despite the remediation work carried out, the survey results confirm the active slope movement at the crawling stage.

3 DATA AND METHODS

Sentinel-1 satellite radar images were provided by the European Space Agency through the Copernicus Open Access Hub web service. The data were processed in the period from October 2014 to January 2018. The monitored site is covered by three satellites orbits: two descending paths no. 51 and 124 and one ascending path no. 175. For a more detailed overview of the satellite radar images used, see Table 1 and the graphical representation of the coverage is shown in Figure 2.

From October 2014 to September 2016, only Sentinel-1A data was available every 12 days. From September 2016 to January 2018, Sentinel-1B satellite data could also be retrieved, reducing the data delivery interval to every 6 days.

Both methods have been applied to SARPROZ © software, which works on the MATLAB © software platform. In SBAS processing, an approach was used in which interferograms were generated within consecutive pairs of radar images. See Figure 3 for a graphical comparison of applied datasets.

The phases of the individual pixels of the interferograms arose as the difference of the phases of the corresponding pixels in the main and subframes. In view of the methods used, there has been a situation where the number of interferograms is the same in both cases. For more information on the amount of interferograms used, see Table 2. Generation of interferograms included modified goldstein filtering that allows chosen the window size. A 15x15 pixel window was used. Interferograms processing also included removing the flat terrain phase fringes and removing the topographic phase component. In order to carry removing the topographic phase component, the digital elevation model was first converted into SAR coordinates.

The interferogram pixel phase provided information about the height differences that occurred in the monitored area at the time interval in which the images were generated. Basically, it is possible to record a change that occurred within a period between two satellite radar images. However, all generated interferograms were used to estimate the average rate over the entire monitored period.

The advantage of using more data is that it was possible to estimate atmospheric artifacts in the interferogram phase and by subtracting them partially reduced the influence of the atmosphere on the electromagnetic signal (Kampes, 2006).

In order to characterize the nature of the changes in the site of interest, a set of stable scatterers was required, which were characterized by high phase stability throughout the time interval investigated. For this purpose, a parameter called the amplitude stability index was used. In this treatment, the values of said parameter were in the range of 0.6 to 0.95. The values differ because the partial objective was to obtain an approximately equal set of stable reflectors for all treatments. Looking at Figure 3, it can be seen that the temporal decorrelation of multiple frame pairs in the PSInSAR method reduces the coherence value, so lower

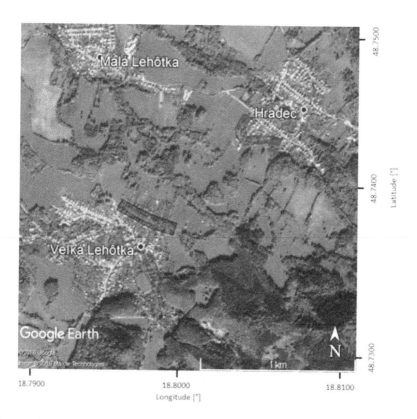

Figure 1.   Monitored area - Hradec and Velka Lehotka. Landslides damaged anthropogenic objects on Pavlovska (red color), Na Staniste (orange color), Remeselnicka (blue color) and Podhorska (purple color) streets.

Table 1.   An overview of the satellite radar images used.

| Orbit | | | | |
Type	Number	Satellite	Number of radar images	Time interval
Ascending	175	Sentinel-1A	89	from 22.11.2014 to 26.01.2018
		Sentinel-1B	32	from 30.09.2016 to 31.10.2017
Descending	51	Sentinel-1A	79	from 09.10.2014 to 21.01.2018
		Sentinel-1B	36	from 04.10.2016 to 27.01.2018
	124	Sentinel-1A	88	from 24.10.2014 to 26.01.2018
		Sentinel-1B	36	from 02.11.2016 to 20.01.2018

amplitude stability index values have been used to generate a plurality of reflection elements. In the SBAS method, the temporal decorrelation is lower, so higher values of the selected parameter have been applied and the number of stable reflectors has been preserved. The average annual velocity was calculated for each stable reflector.

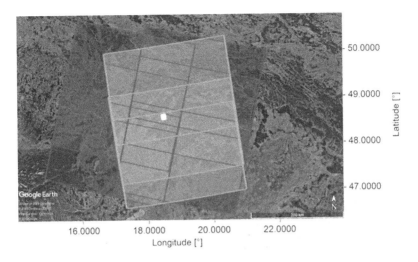

Figure 2.   Area of interest coverage (white rectangle) by radar satellite images: track no. 51 (red), track no. 124 (blue), track no. 175 (green).

Coherence was initially considered when generating a set of stable reflectors. However, coherence is estimated as the average coherence of the reflected signal from all elements in one pixel. That is, the coherence of the dominant element, within the pixels, also includes the decorretion caused most often by the appearance of vegetation at the same pixel. In this way, the pixels located on the boundary between the built-up area and the vegetation cover were mainly affected. Ultimately, this has led to a reduction in the number of generated stable reflectors. By using the amplitude stability index, a larger set is obtained that allows the changes in the selected site to be characterized in more detail.

4   RESULTS AND DISCUSSION

In each stable reflective element, the average velocity of the transmitted signal was estimated in millimeters per year (mm/year). A graphical representation of the resulting average velocities can be found in Figure 4.

The results point to compliance with the findings of engineering-geological surveys and thus, in addition to the northern part, there is movement activity of slopes in other areas of Hradec. It is a larger area than the landslides in 2012 and 2013. In the case of Velka Lehotka, physical activity is detected in the northern and eastern parts of the municipality, which corresponds to the affected areas in 2012 and 2013.

On the graphical outputs (Figure 4), we can see that there are areas in the marginal areas of built-up areas where the presence of stable reflective elements is absent. The likely reason is that, despite the use of the amplitude stability index, vegetation causes strong decoration. A possible way to increase the set in these sections is to apply another SBAS approach, whereby the entire set of satellite radar images is split into smaller sets, and within them, pairs are searched for where the effect of decoration is as small as possible.

In the case of the results of both methods, the speed of movement in the affected parts of municipalities is very similar. These conclusions could lead to the assumption that despite the security work carried out, the site is still active.

An interesting phenomenon is that the results of the processing of radar satellite images from descending tracks no. 51 and 124, the area appears to be declining. With the ascending path, the reflective elements look as if they are rising. This observed phenomenon offers more detailed information on the direction of terrain movement. The Sentinel-1A/B satellites

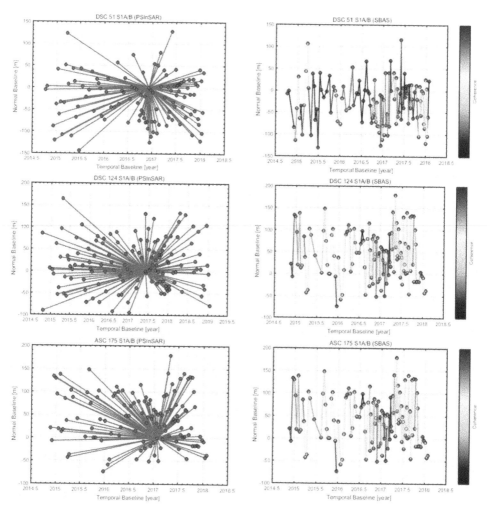

Figure 3. Graphical comparison of frame pairs distribution. The graphs show the dimension of the normal baseline on the vertical axis and the temporal baselines on the horizontal axis. On the left side are graphs from the PSInSAR method. All interferograms have one common image (master) approximately in the middle of the dataset. On the right side are graphs from the SBAS method. Interferograms were created sequentially between consecutive images.

Table 2. Overview of interferograms used.

Orbit			
Type	Number	Method	Number of interferograms
Ascending	175	PSInSAR	120
		SBAS	120
Descending	51	PSInSAR	114
		SBAS	114
	124	PSInSAR	123
		SBAS	123

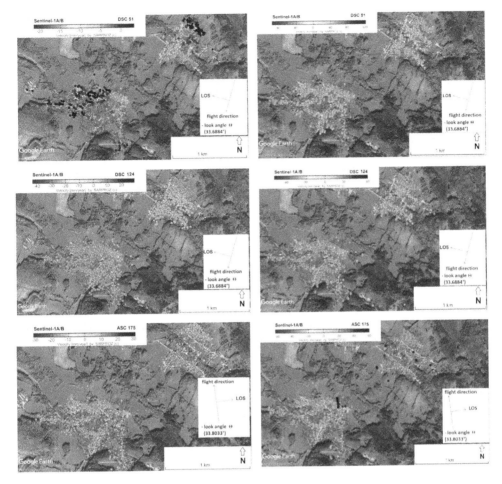

Figure 4. Mean movement velocities of stable reflectors estimated by PSInSAR (left) and SBAS (right).

monitor the Earth's surface from the right side and move along the near polar path. This means that when moving on a descending path, they send a signal approximately from east to west, and in the opposite direction when moving on the ascending path. Based on this information, the idea is that the terrain is moving east to west. Therefore, in ascending path, the terrain appears to rise as it approaches the satellite track. More detailed motion information can be obtained by decomposing the velocity vector in the direction of the transmitted signal to a horizontal and vertical component. The decomposition of PSInSAR processing results from the territory of this paper is presented in (Ondrejka et al., 2016), and along with the time series in (Lesko et al., 2018).

5 CONCLUSION

Both methods, PSInSAR and SBAS, detected slope movement in Hradec and Velka Lehotka. Most persistent scatterers reach velocities in the range of -5 to -20 mm/year in Hradec and -5 to -10 mm/year in Velka Lehotka, in the case of descending orbits. For ascending orbit, there are detected almost identical values but with opposite sign. These values represent the movement of the slope at the crawling stage, which coincides with (Ilkanic, 2013) despite the remediation work in the area. From the estimated data, it can be assumed that the area moves from east to west, which corresponds to the slope of the area and is also confirmed by the decomposition of the velocity vector in (Ondrejka et al., 2016) and (Lesko et al., 2018).

The results are influenced by the decoration caused by vegetation predominantly in peripheral areas. This is manifested by a reduced number of stable reflection elements and an increase in the reflected signal noise of the elements that have met the desired amplitude stability condition. The SBAS method was used t obtain more reflection elements in areas initially covered by vegetation. Interestingly, almost identical numbers of elements were achieved in both methods. However, the opposite effect was noted. Thus, in areas partially covered by vegetation, less reflection elements were found by SBAS than by PSInSAR. This fact is especially visible in the southern part of Velka Lehotka and in the eastern part of Hradec (Figure 4). This paper shows graphical outputs showing approximately the same number of points using a different value of amplitude stability index. For SBAS, a higher value for this parameter was used. This is probably the reason why the number of points has decreased. Using the same parameter value, the number of reflection elements for SBAS is higher.

In both cases, there was a loss of reflective elements at the border between buil-up area and vegetation. This undesirable fact was partially reduced by using amplitude stability index instead of coherence to generate a network of persistent scatterers. In this way, their number was increased.

In small areas, such as the area examined in this article (5x5 km), the effect ot the atmosphere should not be reflected in the results. However, tha master image was chosen so that it came from approximately the middle of the data set and at the same time that weather was suitable on the given date (no clouds, no precipitation, no snow period). Only then did the effect of the atmosphere from the interferometric phase be reduced relative to the state of the atmosphere in the master image.

Determination of geodetic accuracy of results would be possible by comparison with results of geodetic measurements. For this reason, it is necessary to realize leveling of GNSS measurements at the site of interest. The next step to a detailed motion analysis is to split the datasets into individual years.

The radar data used was generated in the C-band of electromagnetic radiation. However, in order to better understand the effect of wavelength, spatial resolution, and repetition period on the estimation of the velocity of the reflective elements, it is also necessary to compare the results with the L and X bands.

ACKNOWLEDGMENT

The Sentinel-1 data was provided by ESA as part of the policy of providing free and full data as part of Copernicus. This work was supported by the Slovak Grant Agency VEGA within the project number 2/0100/20. The authors thank the State Geological Institute of Dionýz Štúr and the Upper Nitra Mine Prievidza, a.s. for discussion and cooperation.

REFERENCES

Shanker P., Casu F., Zebker H. A., Lanari R. "Comparison of Persistent Scatterers and Small Baseline Time-Series InSAR Results: A Case Study of the San Francisco Bay Area", *IEEE Geoscience and Remote Sensing Letters*, vol. 8, no. 4, pp. 592–596, July 2011.

Ferretti A., Prati C., Rocca F. "Permanent Scatterers in SAR Interferometry", *IEEE Trans. Geoscience and Remote Sensing*, vol. 39, no. 1, pp. 8–20, January 2001.

Bernardino P., Fornaro G., Lanari R., Sansosti E. "A New Algorithm for Surface Deformation Monitoring Based on Small Baseline Differential SAR Interferograms", *IEEE Trans. Geoscience and Remote Sensing*, vol. 40, no. 11, pp. 2375–2383, November 2002.

Casu F. "The Small Baseline Subset Technique: Performance Assessment and New Developments fo Surface Deformation Analysis of Very Extend Areas", *XXI Cycle*, February 2009.

Lanari R., Mora O., Manunta M., Mallorquí J. J., Bernardino P., Sansosti E. "A Small-Baseline Approach for Investigating Deformations on Full-Resoution Differential SAR Interferograms", *Geoscience and Remote Sensing*, vol 42, no. 7, pp. 1377–1386, 2004.

Ferretti A., Prati C., Rocca F. "Nonlinear Subsidence Rate Estimation using Permanent Scatterer in Differential SAR Interferometry", *Geoscience and Remote Sensing*, vol. 38, no. 5, pp. 2202–2212, 2000.

Bürgmann R., Hilley G., Ferretti A., Novali F. "Resolving Vertical Tectonics in the San Francisco Bay Area from Permanent Scatterer InSAR and GPS Analysis", *Geology*, vol. 34, no. 3, pp. 221–224, March 2006.

Lanari R., Casu F., Manzo M., Lundgren P. "Application of SBAS-DInSAR Technique to Fault Creep", *Remote Sensing Environmental*, vol. 109, no. 1, pp. 20–208, July 2007.

Casu F., Manzo M., Lanary R. "A Quantitative Assessment Of the SBAS Algorithm Performance from Surface Deformation Retrieval from DInSAR Data", *Remote Sensing Environmental*, vol. 102, no. 3/4, pp. 195–210, June 2006.

Ferretti A., Savio G., Barzaghi R., Borghi A., Musazzi S., Novali F., Prati C., Rocca F. "Submillimeter Accuraccy of InSAR Time Series: Experimental Validation", *IEEE Trnas. Geoscience and Remote Sensing*, vol. 45, no. 5, pp. 1142–1153, May 2007.

Kampes B. M. "Radar Interferometry: Persistent Scatterer Technique", *Dordrecht: Springer*, 2006. ISBN 978-1-4020-4723-7.

Ondrejka P., Bakoň M., Papčo J., Liščák P., Žilka A. "Monitoring aktivity zosuvného územia Prievidza-Hradec", *Geotechnika*, č. 1, s. 3-12, 2016.

Lesko M., Papčo J., Bakoň M., Liscak P. "Monitoring of Natural Hazards in Slovakia by Using of the Satelite Radar Interferometry", *Procedia Computer Science*, vol. 138, pp. 374–381, 2018.

Advances and Trends in Geodesy, Cartography and Geoinformatics II –
Molčíková, Hurčíková, & Blišťan (eds)
© 2020 Taylor & Francis Group, London, ISBN 978-0-367-34651-5

Spacetime curvature on the surface of the Earth

P. Letko

Department of Theoretical Geodesy, Faculty of Civil Engineering, Slovak University of Technology, Bratislava, Slovakia

ABSTRACT: The paper aims to visualize spacetime curvature to better understand its character. It does so by means of Ricci (scalar) curvature that is computed and plotted on the regular surface of the Earth (reference ellipsoid). The metric field of the Newtonian limit of the theory of relativity is taken into account as Earth's gravity is relatively weak (compared to Sun, white dwarf or black hole). In the Newtonian limit, the source of curvature is the time coordinate which is directly dependent on the gravitational potential field generated by the central body as shown in the paper. The potential of the real Earth is a function of position and time, because of heterogeneous mass distribution and geodynamic processes within its body and is therefore defined in spherical harmonics series expansion. This fact is taken into account in the derivation of the scalar curvature formula. This formula is then used to compute curvature of spacetime in regular global grid defined on the surface of the reference ellipsoid. Because of the distribution of Earth's mass, the map of scalar curvature is compared with density inhomogeneities derived from the map of gravity anomalies as well as with the map of continental plates - dependence between these phenomena is demonstrated graphically. The computation is performed using the GrafLab software and recent combined global geopotential model GOCO05c based on GOCE data.

1 INTRODUCTION

Albert Einstein's general theory of relativity (GR) explains gravitational attraction of the bodies via curvature of the four-dimensional spacetime. Metric tensor field is a key quantity describing geometry of spacetime playing the role of gravitational potential in the classical Newtonian theory of gravitation. Information about spacetime curvature is then stored in Riemann-Christoffel tensor and related quantities of Ricci tensor and Ricci curvature which are all functionals of the metric tensor and its derivatives respectively.

GR utilize a concept of four-dimensional spacetime continuum that is dynamic in space and time. This continuum is curved and not „flat", therefore the basic concepts of Euclidean geometry cannot be used. The geometric properties of this spacetime manifold can be described only by using concepts of differential geometry and tensor calculus (for reference see e.g. Lipschutz, 1969). Einstein's field equations read as follows (e.g. Schutz, 1980).

$$R^{\alpha\beta} - 1/2\left(Rg^{\alpha\beta}\right) = \left(8\pi G/c^4\right)T^{\alpha\beta}, \tag{1}$$

where the left side describes geometric properties of spacetime via the Ricci tensor $R^{\alpha\beta}$, Ricci curvature R and metric tensor $g^{\alpha\beta}$ (playing the role of Newtonian potential in GR), while the right side describes the distribution of mass via the stress-energy-momentum tensor $T^{\alpha\beta}$ (here c is the speed of light and G is gravitational constant). These geometric quantities are derived from the Riemann-Christoffel tensor $R_{\alpha\beta\gamma\delta}$ (see e.g. Hobson, 2006) and each of them has its own significance and visual interpretation (or notion) that we will briefly examine following the approach of (Loveridge, 2004).

The Riemann-Christoffel tensor is the only true indicator of the real curvature of a given surface or a manifold. It measures the curvature via the parallel transport of a vector along two
different paths (from point A to point B) on the manifold via the covariant derivative. If two
different vectors result after the parallel transport, the manifold is curved, i.e. if there is
a single non-zero term in the Riemann-Christoffel tensor (out of 4^n terms, n being the dimension of the manifold), the manifold is curved. Its covariant form is defined as (e.g. Hobson,
2006)

$$R_{\alpha\beta\gamma\delta} = 1/2 \left(g_{\beta\gamma,\delta\alpha} - g_{\alpha\gamma,\delta\beta} + g_{\alpha\delta,\gamma\beta} - g_{\beta\delta,\gamma\alpha} \right) - g^{\omega\sigma} \left(\Gamma_{\omega\alpha\gamma}\Gamma_{\sigma\beta\delta} - \Gamma_{\omega\alpha\delta}\Gamma_{\sigma\beta\gamma} \right), \quad (2)$$

where the comma (,) designates a partial derivative, $g_{\alpha\beta}$ ($g^{\alpha\beta}$) is a covariant (contravariant)
metric tensor, $\Gamma_{\sigma\beta\gamma} = [\alpha\beta, \gamma]$ is a Christoffel symbol of first kind and the greek indices run from
1 to N (representing N-dimensional manifold) and standard Einstein summation convention
is used. By many symmetric properties there are only $N^2(N^2-1)/12$ independent terms. As
such, this tensor measures the non-commutativity of the second covariant derivative and rate
of difference between parallel-transport-resulting vectors.

The Ricci tensor and Ricci (scalar) curvature are defined as (e.g. Hobson, 2006)

$$R_{\alpha\beta} = g^{\eta\mu} R_{\mu\alpha\eta\beta} = \Gamma^{\eta}_{\alpha\beta,\eta} - \Gamma^{\eta}_{\alpha\eta,\beta} + \Gamma^{\xi}_{\alpha\beta}\Gamma^{\eta}_{\xi\eta} - \Gamma^{\xi}_{\alpha\eta}\Gamma^{\eta}_{\xi\beta} , \quad (3)$$

$$R = R^{\alpha}_{\alpha} = g^{\alpha\mu} R_{\mu\alpha} = g^{\alpha\mu} \left(\Gamma^{\eta}_{\mu\alpha,\eta} - \Gamma^{\eta}_{\mu\eta,\alpha} + \Gamma^{\xi}_{\mu\alpha}\Gamma^{\eta}_{\xi\eta} - \Gamma^{\xi}_{\mu\eta}\Gamma^{\eta}_{\xi\alpha} \right), \quad (4)$$

where $\Gamma^{\alpha}_{\beta\gamma}$ is the Christoffel symbol of the second kind.

The Ricci curvature (also called scalar curvature) is a trace of a Ricci tensor (see below)
and, in a way, also the generalization of Gaussian curvature used in differential geometry. It
can be used to describe the difference in volume between a „small" geodesic sphere and
a sphere of same radius in euclidean (flat) space. The change of volume may or may not be
isotropic (Figure 1). The sphere is constructed as a set of all points connected to the center by
a geodesic of fixed length. In Figure 1 the blue arrows schematically represent the forces
applied on the spheres, the red ones represent volume differences.

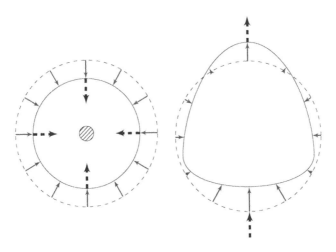

Figure 1. Volume difference between a geodesic sphere (plain line) and an ideal sphere (dashed line) due
to a central mass. There is volume change, so the scalar curvature is not zero. The change of volume is
the same in all directions, so the Ricci tensor is a multiple of the metric tensor and the scalar curvature is
a constant (left); volume difference between a deformed geodesic sphere due to gravity and a perfect
sphere from a flat space that is not isotropic (right). (source: *naturelovesmath.com*).

The Ricci tensor, being a trace of the Riemann-Christoffel tensor, also describes the aforementioned difference of volume, except depending on direction. It measures the curvature via averaging all the sectional curvatures in a particular direction and compares conic sections of this sphere to account for local volume changes and deformation. If we suppose that the volumes of a sphere in flat and curved space were equal, then we would have null Ricci curvature, but a nonzero Ricci tensor because there would be volume changes in different directions (even if they compensate overall) and these changes can be seen by taking the conic sections (Figure 2). A sphere is formed by the set of all points connected to the center by a geodesic of fixed length, in flat space they are straight lines. In a curved space the geodesic lines of fixed length „look shorter or longer" depending on the curvature of space in each direction and make a deformed ball. Scalar curvature vanishes here, so its entire (global) volume is the same. But the conic sections of the same diameter taken in different directions illustrate the change of volume of these sections in these directions and thus curvature of space (Colombano-Rut, 2017).

We focus on the scalar curvature, being a global parameter characterizing curvature of spacetime by single number and thus the most suitable and versatile for its visualization.

3 THE SCALAR CURVATURE FORMULA

The metric tensor must be known if the formula for scalar curvature should be derived. For our purpose the metric tensor of the Newtonian limit of general relativity is used. Newtonian limit describes spacetime curvature in the presence of weak and stationary gravitational fields (metric tensor is a diagonal time-independent matrix) where the objects move with velocities significantly lower than the speed of light c (no distinction between proper and coordinate time τ and t) (for more detail, see e.g. Hobson, 2006). The resulting covariant and contravariant metric tensor in spherical coordinates read (Hofmann-Wellenhof, Moritz, 1993)

$$
\begin{aligned}
g_{\alpha\beta}(x^\alpha) &= \begin{pmatrix} -(1+2V(x^\alpha)/c^2) & 0 & 0 & 0 \\ 0 & 1 & 0 & 0 \\ 0 & 0 & r^2 & 0 \\ 0 & 0 & 0 & r^2 \sin^2\theta \end{pmatrix}, \\
g^{\alpha\beta}(x^\alpha) &= \begin{pmatrix} -(1-2V(x^\alpha)/c^2) & 0 & 0 & 0 \\ 0 & 1 & 0 & 0 \\ 0 & 0 & 1/r^2 & 0 \\ 0 & 0 & 0 & 1/(r^2 \sin^2\theta) \end{pmatrix}.
\end{aligned}
\tag{5}
$$

Here, $V(x^\alpha)$ is gravitational potential given at an event x^α. Because we consider potential of real Earth, it is a function of space and time, even though for our purpose we neglect its temporal variations and is given in terms of spherical harmonic expansion as (e.g. Lowrie, 2011)

Figure 2.   The sphere in a flat (top left) and curved space (bottom left) have the same volume. The conic sections (of the same diameter) made in different directions show the direction-dependent change of volume (top center to bottom right) (source: *naturelovesmath.com*).

$$V(r,\theta,\lambda) \;=\; -GM \,/r \left[1 + \sum_{n=2}^{\infty} (R_0/r)^n \sum_{m=0}^{n} (C_{nm}\, sinm\lambda + S_{nm}\, cosm\lambda)\, P_m(\cos\theta) \right], \qquad (6)$$

where *(-GM/r)* is gravitational potential of a homogeneous sphere, *(r, θ, λ)* spherical coordinates (radial distance, colatitude, longitude), C_{nm}, S_{nm} spherical harmonic coefficients of degree *m* and order *n*, $P_{nm}(cos\theta)$ associated Legendre function of the first kind of degree *m* and order *n*, R_0 radius of the reference sphere of the geopotential model.

Individual nonzero terms of Riemann-Christoffel tensor and Ricci tensor must be computed using (5), (2) and (3) to be able to compute scalar curvature using (4) which is rather tedious. The non-zero Christoffel symbols $\Gamma_{\sigma\beta\gamma}$ and terms of Riemann-Christoffel tensor $R_{\alpha\beta\gamma\delta}$ are (recall that $\Gamma_{\sigma\beta\gamma} = [\alpha\beta, \gamma]$ and we designate $R_{\alpha\beta\gamma\delta} = (\alpha\beta\gamma\delta)$)

$[rt, t], [\theta t, t], [\lambda t, t], [r\theta, \theta], [r\lambda, \lambda], [\theta\theta, r], [\theta\lambda, \lambda], [\lambda\lambda, r], [\lambda\lambda, \theta], [tt, r], [tt, \theta], [tt, \lambda];$
$(rtrt), (\lambda trt), (\lambda t\theta t), (\lambda t\lambda t), (\theta trt), (\theta t\theta t), (\theta t\lambda t), (\lambda r\lambda r), (\lambda\theta\lambda r), (\lambda\theta\lambda\theta).$

Using (3) and (4) the Ricci curvature finally yields

$$R = R^{\alpha}_{\alpha} = g^{\alpha\beta}R_{\alpha\beta} = g^{00}R_{00} + g^{11}R_{11} + g^{22}R_{22} + g^{33}R_{33} = -\left(2/c^2 - 4V/c^4\right)$$
$$\left[V_{,11} + V_{,22}/r^2 + V_{,33}/\left(r^2 sin^2(\theta)\right) + (2/r)V_{,1} + \left(cot(\theta)/r^2\right)V_{,2}\right] + (2/c^4) \qquad (7)$$
$$\left[(V_{,1})^2 + (V_{,2}/r)^2 + (V_{,3}/(rsin(\theta)))^2\right] = -\left(2/c^2 - 4V/c^4\right)\nabla^i\nabla_i V + (2/c^4)g^{ij}\nabla_i\nabla_j V.$$

Here the index *α = 0, 1, 2, 3* designates time and spherical coordinate (*0 = t, 1 = r, 2 = θ, 3 = λ*). Last line represents tensorial form of the formula that can be used in any coordinate system, where ∇_i designates covariant derivative and $\nabla^i\nabla_i$ Laplacian (because $\nabla^i = g^{ij}\nabla_j$).

4 COMPUTATION AND VISUALIZATION OF SPACETIME CURVATURE

Scalar curvature on the surface of the Earth approximated by the reference ellipsoid (GRS80) can be visualized according to (7) but several parameters need to be computed. Parameters of the gravitational field of the real Earth were computed utilizing the global geopotential model (GGM) GOCO05c. This GGM is based on the measurements of GOCE and GRACE satellites and processed by the International Centre for Global Earth Models (ICGEM) of Potsdam (for details see Pail et al., 2017). The parameters (gravitational potential *V*, components of gravitational tensor V_{rr}, $V_{\theta\theta}$, $V_{\lambda\lambda}$ and gravitational acceleration g_r, g_θ, g_λ) were then computed using the spherical harmonic synthesis in GrafLab (GRAvity Field LABoratory) software in MATLAB® software (Bucha, Janák, 2013). The parameters were computed in global regular grid with step of 0,5° (*φ = ± 90°; λ = ± 180°*) using the maximal degree (*n = 720*) of the GGM and visualized using MATLAB-based software Rotating_3d_globe (Bezděk, Sebera, 2013). The map of scalar curvature is shown in Figure 3.

It can be seen from the map that the scalar curvature is negative overall (reaching the value around -1.10^{-25} m^{-2}). But there are also some local characteristics, mainly in the areas of mass inhomogeneities (e.g. mountain ridges – Himalayas, Andes, etc.) as well as borders of the lithospheric plates of either positive or negative values. The scalar curvature is somewhat similar to Gaussian curvature as we have mentioned earlier. Because of the negative overall value, the geometry of the spacetime on the surface of the Earth is hyperbolic (saddle shaped). This correspond to the basic idea of the GR – gravitational attraction is manifestation of the curved spacetime that is pushing the objects in the vicinity of the gravitating (or rather spacecurving) object (in this case the Earth ellipsoid) toward this object. Dependence and

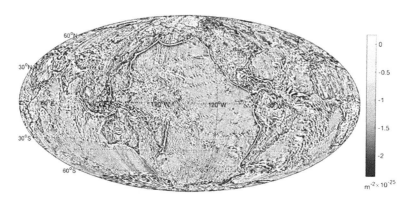

Figure 3. Global map of scalar curvature on the surface of the Earth approximated by reference ellipsoid GRS80.

(GGM: GOCO05c, degree and order $n = m = 720$).

Figure 4. Global map of gravity anomalies on the surface of the Earth approximated by reference ellipsoid GRS80.

(GGM: GOCO05c, degree and order $n = m = 720$.

connection between curvature and mass distribution can be verified. The mass inhomogeneities are manifested by the gravity anomalies (GA) and these were computed using the same GGM and grid. GA were chosen because of their high correlation with terrain; on the other hand, mass causes curvature of spacetime. The map of GA is shown in Fig. The connection is seen particularly good in a detailed part of the Andes mountain area (Figure 5).

The dependence is rather strong and negative, i.e. where there is a weaker gravitational signal (compared to normal gravitation, i.e. GA is negative), the curvature is positive and vice versa – where there is stronger gravitation (GA is positive), the curvature is negative. Regarding the analogy to the Gaussian curvature (see e.g. Lipschutz, 1969), this correlation can be viewed as concave spacetime corresponding to convex terrain (and vice versa). In certain areas there is a step change in the curvature from positive to negative (e.g. areas of colliding Pacific and Australian plates, Himalaya or Andes mountain). Positive curvature corresponds to the spherical geometry, negative to the hyperbolic and null to the flat (as defined in cosmology to describe overall shape of the universe). To verify correlation between the tectonic plate borders and spacetime curvature, the Figure 6 shows the map of tectonic plates based on the Nuvel model data.

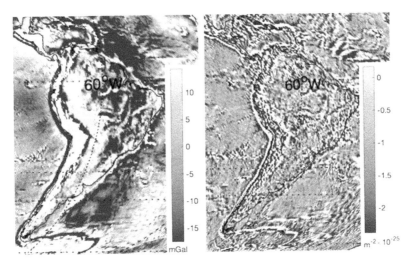

Figure 5.   Detailed view of the Andes mountain – gravity anomaly (on the left) and scalar curvature (on the right).

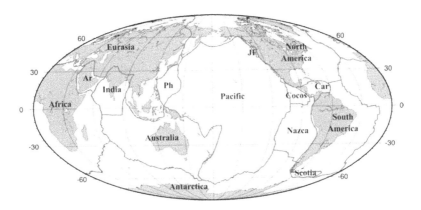

Figure 6.   Map of lithospheric plates (Model: Nuvel) (*source: earthbyte.org*).

The areas of contact between Pacific, Australian and Filipino plate are visible very clearly as well as the North and South America plates with Carribean and Nazca plate. The „Ring of fire" area spanning the border of the Pacific plate is also very definite. The change from positive to negative value in case of GA shows the passage from excess of mass to its absence (regarding the homogeneous distribution of mass in case of normal gravitation) and so from convex terrain to concave. In case of curvature it is change from concave to convex. The Pacific plate moves underneath the Australian plate in its southwest area (this is a convergent boundary) – in this area there is a very distinct interface between negative (east part of the convergent boundary) and positive curvature, while the GA in this area show the opposite character.

5   CONCLUSIONS

The aim of this paper was to visualize spacetime curvature on the surface of the Earth by means of suitable single valued parameter. In Sec.1 the various parameters describing curvature in general theory of relativity are described. Because Riemann-Christoffel tensor encoding the information about real curvature of the spacetime consists of 4^4 terms, the parameter of scalar curvature was chosen. In Sec.2 the formula to compute this curvature was derived based on the Newtonian limit of general theory of relativity and its corresponding metric tensor. The resulting equation include several variables connected to the gravitational field, including gravitational potential, acceleration and gravitational tensor, pointing to a fact, that curvature is real and not just consequence of the utilization of curvilinear coordinates. The results (Sec.3) show connection between curvature of spacetime and inhomogeneous distribution of the Earth masses as well as the tectonic plates. This connection is rather negative meaning that wherever there is an excess of mass, the curvature is negative and where is lack of masses, the curvature is positive, while the overall curvature is negative – equivalence between scalar and Gaussian curvature demonstrates that the overall shape of the spacetime in the vicinity of the Earth (approximated by reference ellipsoid, in this case GRS80) is saddle-shaped corresponding to a hyperbolic geometry, which also leads to a convex/concave relation between shape of the terrain and spacetime.

ACKNOWLEDGEMENTS

This work was supported by project VEGA 1/0750/18.

REFERENCES

Bezděk, A., Sebera, J. (2013): Matlab script for 3D visualizing geodata on a rotating globe, in Computers & Geosciences, Volume 56, 127–130, ISSN 0098-3004, doi: 10.1016/j.cageo.2013.03.007.

Bucha, B., Janák, J. (2013): A MATLAB-based graphical user interface program for computing functionals of the geopotential up to ultra-high degrees and orders. in Computers & Geosciences 56, 186–196, doi:10.1016/j.cageo.2013.03.012.

Colombano-Rut, J. (2017): General relativity loud and clear. online: https://pdfs.semanticscholar.org/bf8c/736a910c6b3db6e2b7f2a77c4d6e19b40eff.pdf

Hobson, M.P., Efstathiou, G., Lasenby, A.N. (2006): General Relativity – An Introduction for Physicists. New York: ISBN 978-0-511-13795-2.

Lipschutz, M. (1969): Schaum's Outline Of Differential Geometry. Mcgraw-Hill: ISBN 978-0-070-37985-5.

Loveridge, L. C. (2004). Physical and geometric interpretations of the Riemann tensor, Ricci tensor, and scalar curvature. arXiv preprint gr-qc/0401099.

Lowrie, W. (2011): A Student's Guide to Geophysical Equations. New York: Cambridge University Press, ISBN 978-0-521-18377-2.

Moritz, H., Hofmann – Wellenhof, B. (1993): Geometry, Relativity, Geodesy, Karlsruhe: Wichmann, ISBN 3-87907-244-2.

Pail R., Gruber, T., Fecher, T. (2017): The Combined Gravity Model GOCO05c, in Surveys of Geophysics, Volume 38, 571–590, ISSN 1573-0956, doi: 10.5880/s10712-016-9406-y.

Schutz, B. (2009): A First Course in General Relativity (2nd edition), New York: Cambridge University Press, ISBN 978-0-521-88705-2.

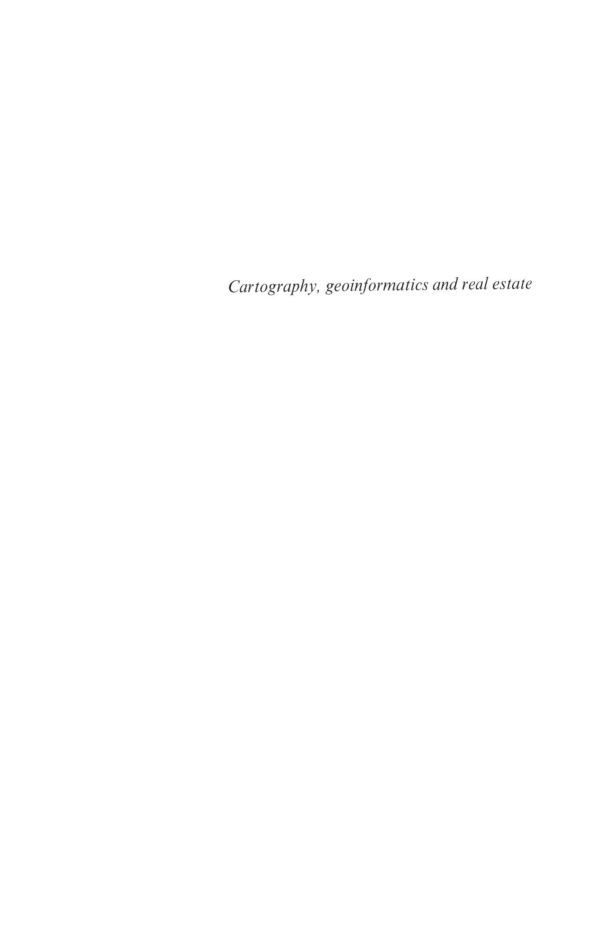

Cartography, geoinformatics and real estate

Advances and Trends in Geodesy, Cartography and Geoinformatics II –
Molčíková, Hurčíková, & Blišťan (eds)
© 2020 Taylor & Francis Group, London, ISBN 978-0-367-34651-5

Interactive creation of tactile maps using geographic information system

M. Andrejka & E. Mičietová
Department of Cartography, Geoinformatics and Remote Sensing, Faculty of Natural Sciences, Comenius University in Bratislava, Bratislava, Slovakia

ABSTRACT: The contribution defines terms in the field of typhlocartography and typhlography. It presents the possibilities of geographic information systems, geoweb and geoinformatic methods used in typhlocartography. The methods of generalization of the theme and geometry of the ZBGIS data model (Primary Base of Geographic Information System) and the Atlas krajiny SR for various degrees of visual impairment are proposed and implemented. Using interactive web geoprocessing – database modelling and generalization of geometry are generated Orientation and Mobility Route Typhlomaps (OMT) for various types of visually impaired people in PDF, JPEG or SVG format for thermoforming Braille printing. The contribution presents a web application and typhlomaps – General Overview or Area Typhlomaps (GOT), Orientation and Mobility Route Typhlomaps (OMT) and Thematic Typhlomaps (TT), which were created by generalization of ZBGIS and selected theme of the Atlas krajiny SR.

1 INTRODUCTION

1.1 *Typhlomaps, typhlocartography, typhlographic*

Typhlomaps present an important element of the haptic perception of geographic information about the location, size, shape and thematic characteristics of the landscape elements. Typhlocartography defines methods of cartographic modelling and cartographic interpretation of positional and thematic component, while using 3 main typhlographic elements: relief point, relief line and relief surface. In typhlography (Myrberg, M. 1978) distinguishes three main forms of expressing reality: a) an iconic representation that describes the appearance of an object using a true, realistic image, b) an analog representation that depicts depth and other aspects of the phenomenon dimension using diagrams, maps, plans, and c) symbolic representation that describes the relationship between the image and the phenomenon through graphs and diagrams.

Haptization – the process of creating geographic information into the form of tactile perception according to (Jesenský, J. 1988) is carried out in two stages: the 1st phase analyzes the structure of the information, the schematization, the 2nd phase is focused on the overall composition and typhlographic elements. (Červenka, P. 1999) who is dealing with the classification of typhlomaps, divides maps according to content, displayed area, scale, way of creation, form of creation, etc. (BANA 2012) specifies two categories: General Overview or Area Maps and Orientation and Mobility Route Maps. The division according to the way of design of typhlomaps depends on the technology and the material that are used for designing typhlomaps (Ojalaet, S. al. 2017). Standardization is important in typhlocartography. Marking geographic objects in Braille is presented by (Chojecka, A. 2012), (Olczyk, M. 2014); (Vondráková, A. 2012) is dealing with 3D typhlomap semiology.

1.2 *Geographic Information System (GIS) and typhlomap creation*

GIS functions have intensified the production capabilities of typhlomaps by a) the availability of digital spatial data sources, b) the possibility of contextual modelling in the geographic database, c) the availability of analytical tools and technologies for cartographic modelling, interpretation and printing using multimedia formats, d) the interoperable Geoweb platform, e) mobile technologies. The current state documents several sources. (Watanabe T. & Yamaguchi T. et al. 2014) implements the creation of 3D typhlomaps according to address points from OpenStreetMap (OpenStreetMap Foundation 2019). Cartographic modelling of 3D typhlomaps in GIS is presented by (Voženílek, V. et al. 2012). Seznam.cz & ELSA (2019) offers download and printing of haptic maps in three scales. Talking Tactile Tablet (Landau, S. & Gourgey, K. 2001, Landau, S. & Wells, L. 2003) technology creates, distributes and displays interactive audio-tactile maps. (Karkkainen, S. 2019) interactively generates typhlomaps from the address and allows to download the file in digital STL format for 3D printer. Polish authors (Mendruń, J. & Olcyzk, M. 2019) enable to access typhloplans of places and tourist points of interest. Mobile applications are using audio messages and vibration signals to navigate visually impaired people on the map currently displayed on the mobile phone screen. (Ciaffoni, G. 2019) by voice support generates a city map by address. The user of the application is still in the middle of the map and the map adjusts to its position, the map navigation is done by vibration signals.

2 RESEARCH OBJECTIVES

Available spatial data and web tools create the opportunities for innovative online automatic creation of typhlomaps (Wabinski, L. & Moscicka, A. 2019). Research objectives of this work are focused on online but personalized creation of typhlomaps (General Overview or Area Typhlomaps (GOT), Orientation and Mobility Route Typhlomaps (OMT) and Thematic Typhlomaps (TT)): in to identification of territory, selection of thematic content and setting of level of spatial generalization.

Specific actual research objectives are: a) identification of official sources of digital spatial data in the Slovak Republic for typhlomaps, b) design and implementation of their generalized data model for GOT, OMT, TT creation in geographic database, c) personalized creation of their cartographic models d) proposal of new cartographic interpretation models for typhlomaps, e) design of 2D and 3D printing models, f) creation of web application for interactive personally configurable cartographic modeling of map content, interactive setting of level of geometry generalization, interactive creation of map signs and print outputs. Interactive personally configuration of typhlomaps is variable for different target groups: for the visu-ally impaired, for people with the residual vision, for people with total blindness and for the assistants who are in daily contact with the blind or with the visually impaired.

3 MATERIAL AND METHODOLOGY

3.1 *Digital data sources*

The basic basis for GIS SR – ZBGIS (ÚGKK 2006) is the national standard of the planimetric and altimetric elements of the SR, representing a digital terrain model in vector form at a scale of 1:10 000. The data model defines the Object Class Catalog (ÚGKK 2017) based on five levels of DIGEST specification (DGIWG 1997): category, subcategory, object class, attributes, and attribute values. The subject of the schematization and design of the Generalized Data Model (GDM) for the GOT and OMT typhlomaps is based on the definitions of the selected ZBGIS elements: AN010 Railway, EC015 Forest, AP050 Pavement, AP030 Road, AL015 Building, AK040 Playground, BH040 Watercourse, EA040 Orchard, garden, AQ140 Car park, CA140 Contour line. Atlas krajiny SR is the source of thematic maps of Slovakia (Hrnčiarová, T. & Abaffi, D. et al. 2002). The subject of schematization of geometry and thematic content for TT were vector layers Average annual air temperature (Chapter 4), Regional towns (Chapter 6), Region boundary (Chapter 2), Villages (Chapter 6).

Panoramic images of mobile mapping have been used to complement the OMT specific elements that are not implemented in ZBGIS, namely pedestrian crossings. Panoramic images were made by system Trimble MX7 with positioning accuracy from 0.02 – 0.05 m to 0.2 – 0.8 m, depending on the quality of GNSS data reception. Object vectorization was performed by Panorama Studio (GISOFT 2019).

3.2 *Geographic database*

ZBGIS and Atlas krajiny SR input data are distributed in digital formats of geographic database *Shapefile* (ESRI 1998). The data models of the adopted topics were modified for typhlomap creation. Generalization was performed – schematization of thematic content and generalization of the geometry of object classes of typhlomaps. Table 1 contains a convert – transformation of the ZBGIS. GDM is implemented in relational database system with support of spatial structure modelling based on 2D topology in accordance with Dimensionally Extended Nine-Intersection Model (Clementini, E. & Di Felice, P.A. 1996).

3.3 *Generalization of graphical typhlomap content*

The methodology of generalization of geometry is based on factors: the distinctive ability of the sight (touch), the scale of the map, the purpose of the map, the character of the territory, the method of cartographic interpretation. The generalization of data model geometry was done according to Pravda (2003) in three steps. Generalization (simplification) is a shape simplification and detail exclusion of the geometry of original. The selection method implements a systematic reduction in the number of objects. The selection may be census or normative. Census selection has determined a minimum dimension condition, is implemented by contextual database demand with regard to both quality and quantity selection. The selection form is based on experimental selection attempts.

The generalized geometries of the map elements are harmonized by methods in the next step – course alignment, object displacement, and exaggeration that is done to improve haptization, if too many objects are concentrated in a small space, or if a selected special object needs to be emphasized. Texts are also the subject of generalization in the map, regarding the limitation of Braille characters as compared to the digital set of text pixel or print text characters. In generalizing a text, we use selection, shift, object matching, and shortened word forms (abbreviation, shortcut words, signs).

Table 1. ZBGIS generalized data model (DIGEST coding).

Object class	Title	Geometry	Attributes **(APPLIED in GDM)**
AN010	Railway	line	**LOC, RGC, RRA, RRC**, ACH, ACV, SOI, CNF
EC015	Forest	area	**TRE**
AP050	Pavement	line	**LOC, TRE**
AP030	Road	line	**RDT, RST, SMC**, WD2
AL015	Building	area	**BFC, CNF**, EXS, **HGT**, HWT, NAM, LOC, TXT
AK040	Playground	area	**USE**
BH140	Watercourse	line	**CNF**, HYC, NAM TUC
EA040	Orchard, garden	area	**CNF**, VEG
AQ140	Car park	area	**CNF**
AT888 (new)	Crosswalk	point	**CNF, TYP**

3.4 *Cartographic modelling of typhlomaps in GIS*

The cartographic model of GOT, OMT, TT typhlomaps are generated in the thesis by two methods: the method of filtering spatial and thematic attributes of geographic objects using geographic data tools and the geoprocessing method, which also uses other tools for generalizing the geometry of map elements. Database relational modelling was applied by SQL-SELECT query with FROM clause and completed by WHERE clause. In the case of database modelling of spatial structures, other geometric operators and functions were applied. Point line and area elements of the typhlomaps were generated by the method of basic attribute queries. A special feature of the GOT map is the 3D georelief model generated by the special interpolation method of the ZBGIS altitude layer.

In the thesis (Andrejka, M., 2019) geoprocessing models for creating GOT and OMT maps are described: a) contextual database modelling of map contents, b) generalization of polygons into compact units according to the setting of minimum size, minimal distance of polygons and simplification of merged polygons, c) generalization of multiple line objects in one layer and line unification according to criteria of smoothing, simplification and spacing setting. The contribution presents cartographic models of OMT and GOT typhlomaps, the content of which consists of the following layers: GOT (AN010, EC015, AP050, AP030, AL015, AK040, BH140, EA040, AQ140, **CA010**), OMT (AN010, EC015, AP050, AP030, AL015, AK040, BH140, EA040, AQ140, **AT888**).

3.5 *Creating map outputs*

Typhlomap outputs were made by printing and using online web application. Printing offset technologies were applied, Braille printer technology, swell-paper and 3D printing. When using a Braille printer in the first step, the text and map elements are converted into Braille characters and are printed, in the second step a chromatic or achromatic offset printing on the map content is performed (used equipment Spot-Dot Emprint). In Swell-paper technology in the first step are printed high-quality black-and-white images, charts or texts on heat-sensitive microcapsule paper by the method of offset printing using standard print formats (e.g. PDF, JPEG). The second step is the insertion of a printed document into a device (technology Zi-Fuse Heater), which, by means of an infrared lamp, produces a relief output from the created offset document on microcapsule paper. The fused deposition modelling (FDM) method was used for 3D printing. The input format for 3D printing of maps was map materials in STL format. To generate them in the GIS environment, a combination of *ArcGIS-ArcMap, ArcScene* (ESRI, 2018) and *Blender* (Community, B. O., 2018) was used.

After the implementation of cartographic map content modelling (*ArcMap, Arc TOOLS, Model Builder*) map layers and geoproces models were published on the map server (*ArcGIS Server*). A web application has been created and published on *ArcGIS Online* (ESRI, 2018) platform using the *APP Builder* server application. The application functionality is original, it allows interactive generation of OMT maps and set generalization of line and polygon map elements according to the target group´s visual impairment. The application generates a print file with output for printing maps using the technology of Swell-paper and Braille printers.

4 RESULTS

4.1 *Typhlomaps TT, OMT and GOT*

Cartographic modelling and the interpretation of the average annual temperature were implemented for thematic typhlomaps. In the maps different scaling, colour and text interpretation are applied. Figure 1 illustrates map outputs for visually impaired people (left), people with residual vision (middle) and blind people (right).

The General Overview and Area Typhlomap in Figure 2 represents (left) the 3D altimetric and planimetric model of the generalized ZBGIS data model. The digital height model in the map was generated from the layer CA010 – Contour line.

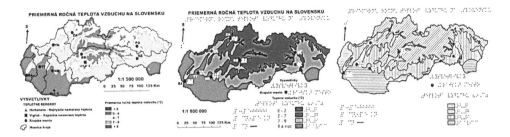

Figure 1. Typhlomaps of average annual air temperature in Slovakia.

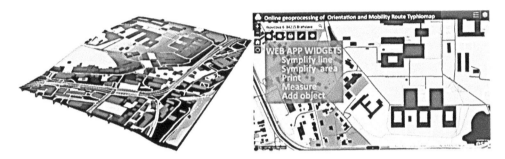

Figure 2.   General overview map: 3D view (left). Web page application window: Online geoprocessing of Orientation and Mobility Route Typhlomap (right).

4.2 Online geoprocessing

Orientation and Mobility Route Maps are generated by a web application with geoprocessing support. The web application style is defined by control elements: a) zoom in (zoom out), b) jump to base position, c) centering, d) measuring of distance and area, e) insertion of geometric shapes and texts, f) simplification of line objects, g) simplification of areal objects, h) printing – allows to print maps in a wide range of digital PDF, JPG (suitable for thermoform printing and offset, SVG (suitable for Braille printing). Figure 2 on the right shows a web application window with controls and map layers. The web application is localized to a URL: https://uvp-uniba.maps.arcgis.com/apps/webappviewer/index.html?id=af0d15a7463949 da8333a61398847baf

5 DISCUSSION

The most of the evaluated existing solutions in the field of automatic (tactile) map generation were described in primary studies from years 2014–2018 (Wabinski, L. & Moscicka, A. 2019). According to this work, most of the identified systems deal with GOT and OMT maps, none of them deal with thematic maps used in education. There is also a problem with the optimization of large spatial databases - the generalization of content and spatial representation. The authors of the work stated that there is no comprehensive solution capable of processing the entire map at once, that existing solutions for automatic tactile map generation are based on raw data. There are no solutions to create automatic thematic tactile maps that could be used in education.

We present a solution that takes into account the needs and recommendations for tactile maps. Thematic generalization of the basic map of Slovakia ZBGIS is presented (Table 1) and generalization of the data model of the selected thematic map (Figure 1). These data models are the original output of this work. In this work is introduced a tool - a web application that allows i) interactive configuration of the orientation map content by selecting topographic elements ZBGIS according to the needs of the target group, ii) interactive personalized setting of spatial generalization of topographic elements for different degrees of visual impairment of the target group, iii) adding custom topographic features to the map; iiii) interactive personalized creation of map signs.

The presented online solution has a methodical character and it is tested on a spatially limited range of input data. The tool for personalized interactive cartographic modelling of orientation maps is original in the context of current online solutions. From the point of view of software flexibility it is suitable to deploy it to the whole territory of the Slovak Republic, as the selection of the area for the creation of the cartographic model of tactile map will always be limited by the scale of the personalized orientation map.

6 CONCLUSION

The presented outputs confirm the importance of the official digital spatial data sources in the SR and the appropriateness of generalization of their data models (thematic and spatial attributes) for the creation of typhlomaps. By their implementation, cartographic modelling, interpretation and distribution using GIS tools, original typhlomap outputs were created. The authors of the work in the GIS environment have developed and presented an original geoinformatic tool – a web application with the support of geoprocessing services for operative online cartographic modelling with the setting of measure of generalization of line and surface planimetric elements.

The possibility of creative and practical use of the achieved results is seen by the authors in the implementation of GIS for the whole territory of the SR, for all topics of the Atlas krajiny SR with the aim of creating digital typhloatlases of the SR. An important extension of the operational possibilities of GIS and cartographic modelling of typhlomaps will be complementing a conceptual GDM and implementation of datasets for GOT and OMT maps by layers of barrier-free objects.

REFERENCES

Andrejka, M. 2019. *GIS as a tool for operative cartographic modelling, interpretation and creation of tactile maps* [Diploma thesis]. Bratislava, The Faculty of Natural Sciences at the Comenius University.

BANA 2012. *Guidelines and Standards for tactile Graphics* [Online] Baltimore, The Braille Authority of North America. Available from: http://www.brailleauthority.org/tg/web-manual/index.html [Accessed. 3th March 2019].

Ciaffoni, G. 2019. *Ariadne GPS. An innovative app for your mobility* [Online] Available from: http://www.ariadnegps.eu/sample-page/ [Accessed. 26th June 2019].

Clementini, E. & Di Felice, P.A. 1996. Model for Representing Topological Relationships Between Complex Geometric Features in Spatial Databases. *Information Sciences* 90(1–4), 121–136.

Community, B. O. 2018. *Blender - a 3D modelling and rendering package. Stichting Blender Foundation,* [Online] Amsterdam. Available from: http://www.blender.org [Accessed. 26th June 2019].

Červenka, P. 1999. *Mapy a orientační plány pro zrakově postižené: metody tvorby a způsoby využití.* Praha: Aula.

DGIWG 1997. ISO/TC 211 19100:1997 *Digital Geographic Information Exchange Standard (DIGEST)* [Online]. Available from: https://www.dgiwg.org/digest/ [accessed. 3th.March 2019].

ESRI 1988. *ESRI Shapefile Technical Description. Redlands,* CA: Environmental Systems Research Institute [Online]. Available from: http://downloads.esri.com/support/whitepapers/mo_/shapefile.pdf [Accessed. 26th.June 2019].

ESRI 2018. Copyrights and Trademarks. [Online]. Available from: https://www.esri.com/en-us/legal/copyright-trademarks [Accessed. 26th.June 2019].

GISOFT 2019. *Panorama studio Software.* [Online]. Available from: http://www.gisoft.cz [Accessed. 3th March 2019].

Hrnčiarová, T. & Abaffi D. et al. 2002. *Atlas krajiny Slovenskej republiky.* Bratislava: Ministerstvo životného prostredia SR a Banská Bystrica: Slovenská agentúra životného prostredia.

Chojecka, A. 2012. *Standardy tworzenia oraz adaptowania map i atlasów dla niewidomych uczniów.* [Online]. Available from: https://tyflomapy.pl/Wprowadzenie_4bc [Accessed. 26th June 2019].

Jesenský J. 1988. *Hmatové vnímání informací s pomocí tyflografiky* Praha: SPN.

Karkkainen, S. 2019. *Touch Mapper* [Online]. Available from:https://touch-mapper.org/[Accessed. 3th March 2019].

Landau, S. & Gourgey, K. 2001. Development of a Talking Tactile Tablet. *Information Technology and Disabilities*, VII (1).

Landau, S. & Wells, L. 2003. Merging tactile sensory input and audio data by means ofthe Talking Tactile Tablet. *Proceedings of EuroHaptic '03*, 414–418.

Mendruń, J & Olcyzk, M. 2019. *Wilno. Plan starego miasta dla newidomych i slabowidziacych* [Online]. Available from: http://tyflomapy.eu/ [Accessed. 26th. June 2019].

Myrberg, M. 1978. Towards an Ergonomic Theory of Text Design and Composition. *Uppsala studies in education Uppsala*, Uppsala University.

Ojala, S., Lahtinen, R. & Hirn, H. 2017. Tactile maps - Finnish O&M instructors' experiences on usability and accessibility. *Finnish Journal of eHealth and eWelfare*. 9(4) 313–321.

Olczyk, M. 2014. *Zasady opracowania map dotykowych dla osób niewidomych i slabowidzących* [Online]. Available from http://ppk.net.pl/artykuly/2014403.pdf [Accessed. 3th March 2019].

OpenStreetMap Foundation 2019. *OpenStreetMap. Open Data Commons Open Database License* [Online]. Available from: https://www.openstreetmap.org/.[Accessed 3th March 2019].

Pravda, J. 2003. *Stručný lexikón kartografie*. Bratislava: VEDA.

Seznam.cz & ELSA 2019. *Haptické mapy* [online] Available from: https://hapticke.mapy.cz/ [Accessed 3th March 2019].

ÚGKK 2006. *Koncepcia tvorby, aktualizácie a správy základnej bázy geografického informačného systému na roky 2006-2010* [Online] Available from: http://www.skgeodesy.sk/files/slovensky/-ugkk/geodezia-kartografia/zb-gis/ktaszbgis06-101.pdf. [Accessed. 3th March 2019].

ÚGKK 2017. *Katalóg tried objektov ZBGIS verzia 2017.8. p. 107.* [Online] Available from: http://www.skgeodesy.sk/files/slovensky/ugkk/geodezia-kartografia/zb-gis/kto_zbgis.pdf [Accessed. 26th .June 2019].

Vondráková, A. 2012. Kartografická sémiologie v moderním typu 3D tyflomap a její vnímání uživateli. *Speciální pedagogika*, 22(1), 12p.

Voženílek, V. et al. 2012. Hypsometry in Tactile Maps. *True - 3D in Cartography* (pp.153–168) [Online]. Available from: http://www.springer.com/gp/book/9783642122712 [Accessed. 3th March 2019].

Wabinski J. &Moscicka A. 2019. Automatic (Tactile) Map Generation – A systematic Literature Review. ISPRS Int. J. Geo-Inf. 2019, 8(7), 293

Watanabe T. & Yamaguchi T. et al. 2014. Tactile Map Automated Creation System Using OpenStreetMap. In: Miesenberger K. & Fels D. et al. *Computers Helping People with Special Needs*.

ICCHP 2014. Lecture Notes in Computer Science, vol 8548. Springer, Cham pp 42–49.

Advances and Trends in Geodesy, Cartography and Geoinformatics II –
Molčíková, Hurčíková, & Blišťan (eds)
© 2020 Taylor & Francis Group, London, ISBN 978-0-367-34651-5

Risk assessment of the parasitozoonoses occurrence using multicriteria analysis approaches

Peter Blišťan, Viera Hurčíkova, Ľudovít Kovanič & Soňa Molčíková
Institute of Geodesy, Cartography and GIS, of Mining, Ecology, Process Control and Geotechnology,
Technical university in Košice, Košice, Slovakia

Ingrid Papajová
Institute of Parasitology SAS, Parasitological Institute, Košice, Slovakia

Monika Blistanova
Faculty of Aeronautics, Technical University of Košice, Košice, Slovakia

ABSTRACT: Parasitic zoonoses pose in the world of current global environmental changes a serious problem not only in developing countries. They occur especially in areas affected by man activities. Specifically, by the negative effect of anthropogenic factors what is altogether exaggerated by poverty level and poor hygiene. In the Slovak Republic, the infections take place primary between the members of marginalised communities. At present we do not have enough data regarding the epidemiology of parasitic diseases in the population living in diverse socio-economic conditions, as well as in animals that live in close proximity. As a consequence location with low hygiene standards represents an important problem regarding the infections epizootology and epidemiology.

The main objective of the present project is in collaboration with results recipient represented by the administration of the Košice self-governing region, where the most of Roma settlements is located, to create a spatial model between the occurrence of most important parasitic diseases and geographical indicators of their development so based on obtained data we will be able to predict the occurrence of parasitic zoonoses in other localities. Results obtained will provide information about the formation of zoonotic infections in Roma settlements, clarify the ways of its dissemination where models of its distribution will be designed. Based on the acquired data a preventive countermeasures reducing the risk of parasitic diseases spread in marginalized groups of people. This project proposal is the first comprehensive study regarding the analysis of risk factors contamination associated with the parasite occurrence and will apply the methodology and approaches of multi criterial evaluation of spatial phenomena into the problematics of parasite zoonoses spread and occurrence.

1 INTRODUCTION

Zoonoses, diseases transmitted by its natural way between man and animals pose a serious health risk. Principally, the zoonoses transmission is accomplished by close contact with domestic animals, especially dogs and cats; which whom we share more than 60 parasitic species. Of about more than 370 parasite species 40 of them are classified as zoonotic. According to the WHO data (2009) more than two billion people are affected by parasitic zoonoses. This is take place not only in developing but also in the industrialised countries, including Slovakia. Parasitic zoonoses represent still a serious problem in the 21st century. According to the WHO, for example, more than 1.2 billion people become infected annually with Ascaris lumbricoides. Additionally, more than 795 million people are parasitized with Trichuris trichiura and 740 million with hookworms.

The prevalence of intestinal parasitic diseases in Slovakia is due to its geographical location and good hygiene conditions relatively low. However, it may be easily dispensable in socially disadvantaged groups of people. Primarily these diseases occur in the population of marginalised communities, which as consequence of various factors are distinct by complete social exclusion. In Slovakia, according to the performed studies and governmentally released strategy papers, the group most at risk and living in poverty which is socially excluded and discriminated is represented by the Roma people. This is in Slovakia a very specific and the most numerous marginalised group.

According to the Atlas of Roma communities (2013) there are 803 Roma settlements in Slovakia. It is home of more than 215 000 people living inside the villages (30.7 %), on the outskirts of the villages (40.4%) or in segregated settlements (16.6 %). Most of the Roma population live with the majority of the population (46.5%) and only 18.4% live in segregated settlements (Filadelfiová et al., 2007; Mušinka et al, 2014). Roma ethnic group inhabit shacks and houses in the villages, apartments with reduced living standards and apartments found on housing estates in urban areas (Dubayová, 2001). Almost 1/3 of these settlements are built illegally without permit. In the cities, we see urban areas which are exclusively inhabited by Roma people and commonly called - Roma ghettos.

The highest percentage of municipalities with Roma community is in Košice self – governing region - KSGR (58.2 %). Also most of the Roma people in this region live in segregated settlements what present the worst living scenario regarding the existence of low personal and municipal hygiene. This is associated with the presence of abandoned waste and high density of rodents and insects around. Very low living standards are linked with poor infrastructure where 20.0 % of settlements have limited access to the main paved roads. Regarding common utilities infrastructure in the Roma settlements the electricity is most available (91.0 % of settlements). The least accessible is public waste management. Only 41.7 % of these settlements have public sewer system what represent the highest number of communities without sewer in Košice region. Gas is unavailable in 16.5 % of settlements and water pipes in 17.4 % of the settlements. Again, the worst situation is in segregated settlements of KSGR, where drinking water is available only in 66.5 % of segregated settlements. Disturbing is the reality that up to 7.2 % of dwellings do not have direct access to the drinking water, Drinking water is often from natural sources such as streams or creeks and consumed on a regular basis without proper sanitary treatment.

2 RISK ASSESSMENT OF THE PARASITOZOONOSES OCCURRENCE

Slovakia does not have an existing comprehensive system for monitoring the health of socially disadvantaged groups of people in effect. The Roma population is the population with the progressive age structure, i.e. with a high proportion of the younger population and with a low proportion of the population over the age of 60. Life expectancy, which is considered a fundamental indicator of the population health status, is in the Roma's men and women only 55.3 or 59.5 years, respectively. WHO specifies the same numbers for the life expectancies in developing countries. The health status of Roma citizens is much worse when compared with the general (major) population. Particularly, there is a high risk of diseases linked with low hygiene standards. The most affected group are children who are often exposed to the environmental and anthropogenic risks. Among of above mentioned parasitoses the soil and man transmitted are the most important. They are represented by ascariasis (A. lumbricoides), trichuriasis (Trichuris trichuris), ancylostomiasis (Ancylostoma duodenale) and necatoriasis (Necator americanus), which occur in a chromic form especially in children living in poverty. While ancylostomiasis and necatoriasis are absent in our territory and considered imported diseases the ascariasis and trichuriasis are typical for our territory, particularly in areas with low hygiene standards. This was confirmed in our pilot study where the occurrence of soil transmitted helminthoses was examined in randomly selected children living in eight segregated settlements located in Košice and Prešov region. We have detected helminths eggs in 21.3 % of randomly examined Roma children at age 0-14. The predominant eggs found were: A. lumbricoides, T. trichiura, Hymenolepis nana, and H. diminuta where an unapparent chronic infection outcome is suggested (Pipiková and Papajová, 2015; Papajová et al., 2016). 24.7 % of children from marginalised Rome settlements have been diagnosed with giardiasis. Children of the major population were only randomly infected. The settlements environment where children lived and played was contaminated extensively with the parasitic propagative stages, where up to 81.6% of examined soil samples were contaminated with helminthic eggs (Ascaris spp., Toxocara spp., Trichuris spp., and strongyloid eggs). Occasionally the eggs of Toxascaris leonina and Spirocerca lupi were detected. The highest soil contamination was detected in villages Jarovnice, Zemplínska Teplica and Sečovce. Frequent eggs findings in the soil correlated with the positivity of examined dogs which shared the same living space and were able to move around freely. 84.6 % of examined dogs living in the settlements were positive for helminthic eggs. The highest-prevalence was associated with the presence of T. canis, T. leonina, T. vulpis, Capillaria spp., and Ancylostoma spp. eggs (Papajová et al., 2014; Pipiková et al., 2017b). Alarming is fact that up to half of the examined dogs excrement samples contained eggs of ascarids and canis origin, the causative agent of severe zoonotic disease - human toxocarosis.

Geographic informatics data system (GIS) based on analysis of variable factors such as soil, climate, temperature, annual rainfall, land exploitation represent a very complex approach contributing to the analyses of local parasitic zoonoses occurrence. At this point during the analysis and identification of risk zones, the method of multi-criteria analysis will be utilized. So far in parasitology this novel approach was used only for malaria epidemiology research and mapping of risk zones where mosquitoes are main contributors of disease spread. Another excellent example of this approach is study on the effect of annual climate changes on the intensity of trematodes transmission.

In the context of above-presented data the main goal of the submitted project is to address an extremely important issue that is the definition of the risk factors, and determine the development and spread of parasitic diseases in people living in contaminated environment of Košice self-governing region. Novelty and originality of this project is the multi-criteria analysis and spatial modelling between the occurrence of the most important parasitic diseases and geographical indicators of their development. Based on these results we will create and compute various models of parasitic diseases spread which will be utilized within Kosice self-governing region. We expect that this original approach will be used by local governments in other Slovakian regions. The objective is to provide solutions for the improvement the life of people living in marginalized communities. Since there is lack of data about the medical and veterinary health status, and we are missing some essential information about the contaminated environment, the realization of such study is imminent and necessary. Our pilot studies already provided basis for this project proposal. It should be stressed, that under certain circumstances, the citizens belonging to the marginalized groups may become a source of infection for the major population. The level of the acquired knowledge in a given topic urges us to continue in a search for better actions to protect the environment against the infections caused by the parasitic developmental and propagative stages. The outcomes of the proposed research will lead to the improvement of the minor and major population health within studied territory. The proposed project will also advance knowledge in the subject matter area home and abroad as well as increase the general awareness level between our citizens and experts on this topic.

The main objectives of the presented project are:

1. Identification of serious parasitic diseases, its incidence and prevalence in minor and major populations in the Košice self-governing region (KSGR). The causality between the pollution exposure and anthropogenic environmental risks of the exposed population will be determined.

2. The analysis of the occurrence of serious animal zoonotic diseases (domestic, residential and wild) as the source of the infection for the marginalized groups and major population in selected localities of the KSGR.
3. Occurrence of the hygienically important propagative stages of endoparasites in the abiotic environment of the KSGR, and analysis of parasites circulation in the environment including physical - chemical analysis of particular components.
4. Utilization of geographic informatics system (GIS), data collection and analysis with subsequent formation of spatial model between the occurrence of the most important parasitic zoonotic diseases. Analyses will connect geographical and climate factors with diseases spread in man (urban and rural settlements) and animals (domestic, wildlife) living in KSGR.
5. Methodology development for the identification and multi criterial evaluation of selected geographical indicators associated with the occurrence of parasitic diseases. The atlas of geographic factors related to the important parasitic diseases within KSGR will be prepared and presented.
6. Application and dissemination of obtained results as well the proposals for the prophylactic and organizational countermeasures for elimination of the spread and potential risks of important parasitic zoonoses for animals and public health in high-risk areas. Presentation of the results in forms of seminars and workshops to the local government represented by KSGR, pediatricians, state veterinary service, local community centers and to the general public.

3 CONCLUSION

The problematics of spatial analyses with exploitation of geographic indicators regarding the spread of parasitic diseases in inhabitants of Kosice self-governing region will be studied by the experts of our collaborator working at the Faculty of Mining, Ecology, Process control and Geotechnologies from Technical University in Kosice (BERG TUKE). For this analyses the GIS instruments such as spatial diameters, standardized elipsoid deviations, concentrations of phenomena, spatial correlations with multi criterial analysis (MCA) and processes of analytical hierarchy (AHP) will be used.

The expert background and practical experience of the project team and its partners guarantees that all project goals will be met and accomplished. This is based particularly on the state-of-the-art experience of the applicants involved: The Institute of Parasitology of SAS and the collaborating institution, BERG TUKE in Košice. Both institutions are equipped with all necessary laboratory instrumentation and have significant experience with research on various scientific assignments. The project outcomes are clearly defined and will integrate broad knowledge and experience of all team members what include parasitology, human and veterinary medicine, environmental ecology, toxicology, analytical chemistry, geoinformatics and GIS. There is optimal combination between human potential and technical resources available to satisfactory elucidate the project goals. The project will be supported also by collaborations and consultations with partner institutions home and abroad (Biology Centre, CAS, v.v.i., České Budějovice; University of Freiburg, Bad Krozingen, Germany; University of Debrecen, Miskolc, Hungary; National Research Institute of Science and Technology for Environment and Agriculture (IRSTEA), Rennes, France; Department of Public Health and Infectious Diseases, University of Rome „La Sapienza", Rome, Italy).

ACKNOWLEDGEMENT

This paper was written thanks to supporting from project APVV-18-0351: Risk assessment of the parasitozoonoses occurrence using multicriteria analysis approaches.

REFERENCES

Dubayová, M., 2001: Rómovia v procesoch kultúrnej zmeny. Prešov: Filozofická fakulta Prešovskej univerzity. 183 s.

Filadelová, J., Daniel, G., & Škobla, D., 2007. Report on the living conditions of Roma in Slovakia. Bratislava: UNDP, Regional Bureau for Europe and CIS.

Mušinka, A., Škobla, D., Hurrle, J., Matlovičová, K., Kling, J., 2014. Atlas rómskych komunít na Slovensku 2013. Bratislava: UNDP, 120 p.

Papajová, I., Pipiková, J., Šoltys, J., Sasáková, N., Takáčová, D., Gregorová, G., 2016: Environment in low hygienic standards settlements in Prešov Region - a potential source of zoonoses. In Congress proceedings from the 5th annual scientific congress on Zoonoses, Foodborne and Waterborne Diseases – Protection of Public and Animal Health. National Focal Point of Slovak Republic for scientific and technical matters for EFSA (ed.). - Bratislava: Ministry of Agriculture and Rural Development of the Slovak Republic, s. 238–240.

Pipiková, J., Papajová, I. 2015: Výskyt zárodkov endoparazitov na verejných priestranstvách v sledovaných rekreačných lokalitách Grécka, Bulharska a Talianska [Occurence of endoparasitic germs in public places in popular tourist destinations in Greece, Bulgaria and Italy]. In Slovenský veterinársky časopis, 2015, roč. 40, č. 3–4, s. 180–183.

Advances and Trends in Geodesy, Cartography and Geoinformatics II –
Molčíková, Hurčíková, & Blišťan (eds)
© 2020 Taylor & Francis Group, London, ISBN 978-0-367-34651-5

Evaluation of spatial distribution of occurrence of parasitozoonoses by GIS tools

Peter Blišťan, Soňa Molčíková, Viera Hurčíkova & Ľudovít Kovanič
Institute of Geodesy, Cartography and GIS, of Mining, Ecology, Process Control and Geotechnology,
Technical university in Košice, Košice, Slovakia

Ingrid Papajová & Júlia Bystrianska
Institute of Parasitology SAS, Parasitological Institute, Košice, Slovakia

ABSTRACT: Geographic information systems are rarely used as a tool to evaluate spatial distribution of phenomena in parasitology in Slovakia. This is the topic of the applied research project - "Risk assessment of the parasitozoonoses occurrence using multicriteria analysis approaches". Its aim is to analyse and evaluate spatial dependencies in the distribution of selected parasites using GIS tools. This paper presents some basic GIS tools suitable for modelling spatial distribution of parasites and evaluating of randomness, as well as GIS tools for cartographic display of these phenomena. Parasite occurrence data are basically point or area data, and these two data groups are used to demonstrate the application of GIS tools.

1 INTRODUCTION

Geographic Information Systems (GIS) is a versatile tool that can be effectively used in specific areas, such as flood modelling (Blistanova et al., 2016), crime analysis, including the assessment of spatial distribution of phenomena, such as occurrence of parasites. Spatial analyses represent a set of techniques developed in different fields to analyse data with an emphasis on spatial relationships of the data. Statistics plays an important role here but procedures have also been derived in geography, econometrics, land use planning, and urban planning. These methods of data analysis are also used on a wider scale, for example, in healthcare – in parasitology. Spatial analyses can be defined as a set of techniques for the analysis and modelling of localized objects where the results of analyses depend on the spatial arrangement of these objects and their properties. According to Horák, it is important to distinguish between spatial analysis of data and analysis of spatial data because the spatial component of data is not always used in analyses (Horák, 2015). There are several types of spatial analysis methods used which are further classified based on applied techniques, data processing methods and spatial presentation type. As regards the applied techniques, spatial analyses are subdivided into statistical (spatial statistics), mapping, mathematical modelling, interpolation, localization and allocation analyses (location of objects, e.g. industrial zone location analysis), network analyses (focusing on transport, technical or natural networks), and other analyses of surroundings and relationship (Horák, 2015).

2 GIS TOOLS FOR EVALUATION OF SPATIAL DISTRIBUTION OF PARASITES

2.1 *Spatial statistics methods for point data*

Descriptive statistical methods are used in GIS to determine the position characteristics and dispersion characteristics of spatial point data. These statistical methods describe the distribution of points using basic statistical characteristics, such as geographic centre, median, standard deviation ellipse, etc. The aim of these methods is to compare multiple sets of point data sets for the purpose of, for instance, monitoring the progress in time and space. Thus, the basic characteristics of each spatial data set is the average geographic centre, which is the centre of the point group lying at the point of the X and Y coordinates average. The average centre has similar problems as the arithmetic mean – predominantly, it is its sensitivity to extreme (outlying) points but also to clusters of points. The nature of dispersion of points is well described by the standard deviation that corresponds with the dispersion measurements around the mean centre. The spatial representation of standard deviation is, in the case of data anisotropy, an ellipse of the standard deviation. It will show possible anisotropy and its spatial characteristics. Based on its size (the ratio of the main to the secondary axis), one can determine the significance of anisotropy - the higher the ratio, the more pronounced anisotropy is; based on the rotation of half-axes towards the coordinate system the positions of the most important outlying observation groups affecting the average centre can also be determined.

2.2 *Hot Spot analyses and transformation of point data into continuous surfaces*

In general, we can state that Hot Spot are places where high values are clustering. The opposite of Hot Spots is Cold Spots which we define as places where low values are clustering. There are currently several techniques for Hot Spot analyses, and single tools within the frame of spatial analyses are also available. The most commonly used methods include the so-called standard deviation ellipses (Spatial Ellipses), thematic mapping of territories, quadrant method and estimates with kernel density function (Chainey et al., 2008).

Hot spot analysis demonstrates one form of analysis of the concentration of phenomena in space. Hot spot analysis calculates Gegis-Ord Gi * statistics for each object. The result of the calculation is the Z-score and P-value, on the basis of which hot and cold spots are calculated. Z-score is a test of statistical significance that helps to decide whether or not to accept the null hypothesis which states: "Values (elements) are distributed randomly and thus are not spatial clustering". P-value is a probability. A high value of Z-score and a low P-value indicate the presence of a spatial cluster of high values in the neighbourhood - Hot Spot. A low negative value of Z-score and a small P-value for a given element prove the existence of a spatial cluster of low values in the neighbourhood - Cold Spot. The larger the Z-score, the more intense the clustering. The Z-score close to zero means that there is no cluster (Kelemen et al., 2015).

The tool compares objects and their values with neighbouring objects and their values. A high-value object is interesting but may not be a statistically significant Hot Spot. To be statistically significant, an object must not only have a high value but it must also be surrounded by objects with high values.

The quadrant method builds on monitoring the frequency of events in defined cells. Various methods with regular and irregular grids are used for transformation. The number of events falling into individual cells indicates the value of continuous surface in a given space (Horák, 2015). The principle is shown in Figure 1 below.

Cells showing high values of the observed phenomenon and identified clusters can be marked as Hot Spot. The utilization of the results of the quadrant method is dependent on the determination of cell size which predetermines the spatial resolution of the analysis. Kernel Density Estimation is a method of estimation that analyses the intensity of the observed phenomenon using a continuous field. The analysis evaluates the distance and statistical significance of individual points to the neighbourhood depending on the set distance of the band. A fine grid is created above the field of points and in each grid point, the contribution of each event is calculated using the kernel estimation function (Figure 2). An overlap is calculated for each cell in the grid (raster) this way quantifying the significance of the phenomenon at certain locations in the studied area (Horák, 2015). With this method, it is possible to select either global Hot Spot or local Hot Spot. Hot Spot analysis methods identify areas at high risk based on localized data.

Figure 1.   The principle of quadrate method of transformation of point data into continuous field.

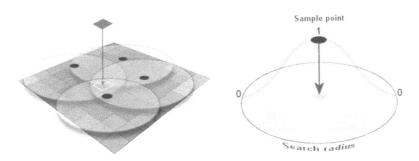

Figure 2.   The principle of the method of kernel density estimation.

2.3   *Methods of cartographic visualization of data in GIS*

This term denotes the selection of map symbols and the way of their distribution in the map field. GIS systems make it possible to create several types of basic thematic map outputs. The choice of the appropriate type depends on the content, purpose, input data and the character of the objects and phenomena displayed (Blišťan, Rapant, 2013). The most commonly used methods of mapping include:

- *method of point symbol,*
- *method of line symbol,*
- *method of isocurve symbol - it is a method of continuous curves with constant value ex-pressing continuity of occurrence of quantitative characteristics of a certain phenomenon, e.g. altitude - isohypses, pressure - isobars, temperature - isotherms, etc.,*
- *method of cartogram,*

- *method of cartodiagram,*
- *cartographic anamorphosis - this is a special method that is rarely used in practical cartography as it changes the overall geometric design of the map and thus changes the appearance of the map.*

2.3.1 *Point character method*
The point character method, as the name implies, can be used on a point layer. It is particularly suitable for displaying the position and quality of discontinuous point objects or phenomena. Symbol size or colour gradation can also reflect the quantitative aspect of a phenomenon that changes discontinuously or corresponds to a range of values (Krtička, Adamec, Bednar, 2012). An example of use is shown in Figure 3 where parasite occurrence and their absolute number is presented.

2.3.2 *Line character method*
The line character method expresses the character, direction and length of line objects, such as roads or railways. It can have the character of movement lines, e.g. wind directions, direction of movement of troops, etc., or the character of diagram lines, e.g. quantity and direction of transport of goods expressed in line thickness etc.

2.3.3 *Method of cartogram*
The method of cartograms is the most used means of expression of thematic cartography. The cartogram shows statistical data calculated on relative values on predefined territorial units.

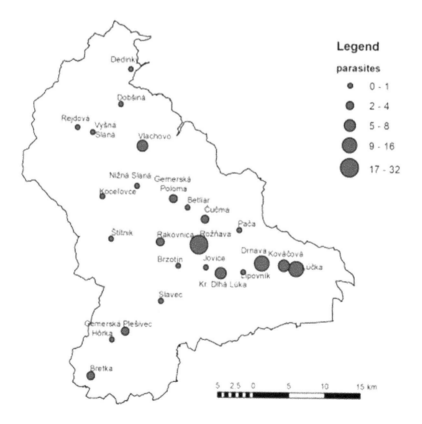

Figure 3. Using the point mark method to display the occurrence of parasites in Rožňava district.

162

The most frequently chosen territorial units are:

- *geographical boundaries - these may be administrative units (regions, districts, etc.),*
- *physical areas (geomorphological units, etc.) or economic zones (agglomerations, catchment areas, etc.),*
- *geometric boundaries - these can be sides of a square, triangle, hexagon, etc. (Krticka, Adamec, Bednar, 2012).*

In geographical practice, cartograms with geographic boundaries predominate, but the use of cartograms with geometric boundaries has also many advantages, for example, when changing the system of administrative divisions. The characteristic feature of cartograms is that the observed phenomenon occurs in the given territory unevenly and is expressed here by a mean value or an interval. It is important to correctly identify the individual intervals – categories, since their number and method of classification has a major impact on the final design of the cartogram. GIS systems offer several methods for classifying categories, namely manual, even-interval, defined-interval, quantiles, natural breaks, geometric intervals and standard deviation.

Their selection depends on the nature of data (Krtička, Adamec, Bednář, 2012). From a methodological point of view, it is important that the calculation refers to the specific unit of area (e.g. population density - inhabitants per square kilometre, grain yield in tonnes per hectare). If the calculation is related to a unit different from the areal unit, it is a pseudo-cartogram, e.g. number of parasites/number of inhabitants (Figure 4).

2.3.4 *Method of cartodiagram*

Cartodiagram is a map in which the statistical data for respective territorial units are marked using diagrams. The expressed values are always in absolute form. The diagram is usually placed in the centre of the area. Diagrams can be point, line and area (Kaňok, 1999). Modern GIS systems offer a choice of different types of diagrams, e.g. column, pie and proportional. They also offer the option of converting the display to selected units.

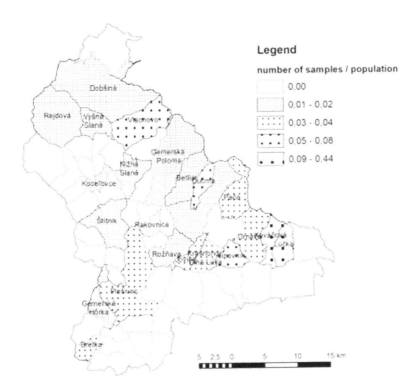

Figure 4.   Use of cartograms to show the number of taken samples in Rožňava district.

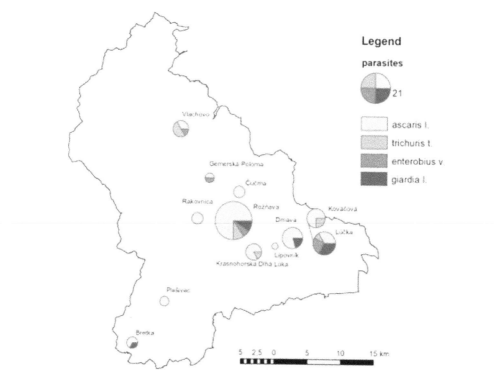

Figure 5.   Use of cartodiagrams to show the occurrence of parasites in Rožňava district.

Figure 5 Shows the number of parasites in Rožňava district with the expression of parasite species using a pie chart.

2.4   *Software*

Spatial data were processed using MS Excel spreadsheet tools, Microsoft Access database system and Esri Geographic Information System called ArcGIS. ArcGIS is known as a universal tool for processing and presentation of spatial data and is used in various fields. ArcGIS Desktop is a basic tool for creating maps, editing data, acquiring information from maps for basic map analyses and modelling. ArcGIS functionality can be expanded through extensions, such as Spatial Analyst, 3D analyst, Geostatistical analyst, etc. Specifically, statistical tools within analytical tools, conversion tools, spatial statistics tools and visualization tools have been used to achieve this goal.

3   CONCLUSION

GIS tools such as analytic tools, spatial analyses and visualization tools are applicable to different types of spatial data. Spatial analyses were also applied to evaluation of spatial occurrences of parasites within the research project "Risk assessment of the parasitozoonoses occurrence using multi-criteria analysis approaches". The aim of the paper was to present basic types of analyses and imaging tools of GIS in processing of point and area data about the occurrence of parasites using GIS tools.

ACKNOWLEDGEMENT

This paper was written thanks to supporting from project APVV-18-0351: Risk assessment of the parasitozoonoses occurrence using multicriteria analysis approaches.

REFERENCES

Blišťan, P., Rapant, P., 2013. Geografické informačné systémy I. Úvod do problematiky a základné pojmy. Vysokoškolská učebnica. Edičné stredisko AMS, F BERG. 107 str.

Blistanova, M., Zeleňáková, M., Blistan, P. and Ferencz, V., Assessment of flood vulnerability in Bodva river basin, Slovakia. Acta Montanistica Slovaca. Volo. 21, no. 1 (2016), p. 19–28.

Horák, J., 2015. Prostorové analýzy dat, VŠB-TU Ostrava, Institut geoinformatiky, 6. vydání, 2015, Ostrava.

Chainey S., Tompson L., Uhlig S. The utility of hotspot mapping for predicting spatial patterns of crime. In: Security Journal Vol.21. 2008. pp 4–28.

Juhásová, L., Králová-Hromadová, I, Zeleňáková, M., Blišťan, P., Bazsalovicsová, E., Transmission risk assessment of invasive fluke Fascioloides magna using GIS-modelling and multicriteria analysis methods. Helminthologia. Vol. 54, no. 2 (2017), p. 119–131.

Kaňok, J., 1999. Klasifikace stupnic a zásady jejich tvorby pro kartogram a kartodiagram. Kartografické listy, č.7, 1999. 75–86.

Kelemen, M., Križovský, S., Blišťanová, M., Blišťan, P., Kováčová, L. Vplyv kamerového systému na priestupkovosť v meste Košice. Bezpečnostná veda a vzdelávanie ako faktory účinnej prevencie), Košice: Vysoká škola bezpečnostného manažérstva v Košiciach, 2015, s.111. ISBN: 978–80–8185–005–9.

Krtička, L., Adamec, M., Bednář, P., 2012. Manuál pracovních postupů v GIS pro oblast sociálního výzkumu a sociální práci. Ostravská univerzita v Ostravě, 1. vydanie, 2012, ISBN 978-80-7464-155-8, 147 str.

Advances and Trends in Geodesy, Cartography and Geoinformatics II –
Molčíková, Hurčíková, & Blišťan (eds)
© 2020 Taylor & Francis Group, London, ISBN 978-0-367-34651-5

GIS-based risk assessment - mapping risk

Monika Blistanová
Faculty of Aeronautics, Technical University of Košice, Košice, Slovakia

Peter Blišťan
Institute of Geodesy, Cartography and GIS, of Mining, Ecology, Process Control and Geotechnology, Technical university in Košice, Košice, Slovakia

ABSTRACT: Knowledge of security risks is an important prerequisite for its managing, whether through preventive measures or already with rescue works. Constantly expanding list of potential risks leads to the need to develop methodologies for their evaluation that would be helpful not only in assessing the level of risk. The paper aims to approach the issue of risk mapping as a result of risk assessment in the GIS environment. Risk mapping is already used in all areas, whether in the field of natural hazards, epidemiology, criminology, or technical risks, and its use is extensive. Risk mapping should be based on risk definition, but problem is a number of definitions and diversity of interpretation of terms risk and hazard. A wide variety of methods for assessing hazard and risk are available. Several institutions and scientific societies have proposed guidelines for the preparation of risk maps especially natural. Some of them are already part of the legislative standards. Despite mentioned is currently not established uniform terminology or risk mapping process as it is for risk assessment.

1 INTRODUCTION

Risk is a normal part of life. Knowing the risks and its components is a prerequisite for their management. Risk assessment is the central components of the general process, which furthermore identifies the capacities and resources available to reduce the levels of risk. Geospatial information is critical for all phases of the disaster management cycle, from supporting risk identification and reduction, and developing preparedness solutions, to prioritizing response and recovery efforts *(Albano, Sole, 2018)*. Numerous scientific studies have introduced several methods on how to assess and map risk. These studies focused on risk related to various processes at local or regional levels. However, all these studies require a significant amount of detailed data, adequate technology, and experts to develop and run complex models. *(Papathoma – Kohle et al., 2016)* Although the risk is the focus of many regulations, recommendations, and directives at national, and international levels, risk assessment studies, there are no clearly defined terms or procedures for risk assessment based on spatial data. The understanding the concept of risk and its components is the subject of this paper.

2 RISK - THEORETICAL VIEW

According to Aven (2016), the risk field has two main tasks. First is to use risk assessments and risk management to study the risk. The second task is to perform generic risk research and development, related to concepts, theories, frameworks, approaches, principles, methods, and models to understand, assess, characterize, communicate and (in a broad sense) manage risks.

There are substantial differences in the way of risk is defined, understood, analyzed, and measured in various sectors. Risk is a combination of the consequences of an event (hazard) and the associated likelihood/probability of its occurrence *(ISO 31 010)*. According to the definition, risk can be described:

$$\text{Risk} = \text{hazard impact x probability of occurrence} \qquad (1)$$

A hazard is a dangerous phenomenon, substance, human activity or condition that may cause loss of life, injury or other health impacts, property damage, loss of livelihoods and services, social and economic disruption, or environmental damage *(Ciurean et al., 2013)*.

Risk should neither be defined nor managed without placing it in a cultural, sociological, and psychological context, especially disaster risks. It should be clarified in what ways the various perspectives influence the way we look at risk. *(Hokstad, Steiro, 2006)*. Impacts of the environment are often expressed in terms of vulnerability. The current literature encompasses more than 25 different definitions, concepts, and methods to systematize vulnerability *(Birkmann, 2006)*. One of the best-known definitions was formulated by the International Strategy for Disaster Reduction, which defines vulnerability as *the conditions determined by physical, social, economic and environmental factors or processes, which increase the susceptibility of a community to the impact of hazards. (UNISDR, 2002)* Mathematically, the risks as a function of hazard and vulnerability can be formulated as follows:

$$\text{Risk} = \text{Hazard x Vulnerability} \qquad (2)$$

There are generally four factors (or dimensions) of vulnerability *(UNISDR, 2004)*:

- *The physical/functional - may be determined by aspects such as density levels, the remoteness of a settlement, its siting, design, and materials used for critical infrastructure and housing.*

- *The economic aspects -. Links between the eradication of poverty, impact consequences on recovery conditions from natural disasters, and the state of the environmental resource base upon which both depend are crucial. Inadequate access to critical and socio-economic infrastructure, including communication networks, utilities and supplies, transportation, water, sewage, and health care facilities, increase people's exposure to risk.*

- *The social aspects - includes aspects related to levels of literacy and education, the existence of peace and security, access to human rights, systems of good governance, social equity, traditional positive values, knowledge structures, customs, and ideological beliefs, and overall collective organizational systems. Organizational and management structures play an essential role in the level of social vulnerability.*

- *The environmental aspects - covers an extensive range of issues in the interacting social, economic and ecological aspects of sustainable development as it relates to disaster risk reduction, such as the state of resource degradation, loss of resilience of the environmental systems, loss of biodiversity, exposure to toxic and hazardous pollutants.*

Ciurean et al. (2013) define the fifth aspect/dimension:

- *The political/institutional - refers to those political or institutional actions, e.g., livelihood diversification, risk mitigation strategies, regulation control, etc., or characteristics that determine differential coping capacities and exposure to hazards and associated impacts.*

In addition to hazard and vulnerability factors, the factor capacity plays a role in increasing or decreasing risk, especially disaster risk *(Sari, et al. 2017)*. Capacity can be defined as a combination of all the strengths and resources available within a community or organization that can reduce the level of risk or the effects of a disaster. The capacity expresses the suitability to "stand against disaster." Mathematically can by a relationship describe:

$$Risk \ = \ Hazard \ x \ Vulnerability/ \ Capacity \qquad (3)$$

3 GIS BASED RISK ASSESSMENT - MAPPING RISK

During recent years, the improvement of geographic information systems (GIS) has allowed us to consider risks by including spatial information. The result of the risk assessment in GIS systems - risk maps display levels of risk across a geographical area.

Based on the above definitions and mathematical relationships, both hazard and vulnerability (impact) factors need to be assessed for risk assessment. The objective of a hazard assessment is to identify the probability of occurrence, in a specified future period, as well as its intensity and area of impact *(UNISDR, 2004)*. Information regarding the hazard and the impact is usually based on historical data and the documentation on of past events. Historical events are often used in deterministic analyses that assess the impact of recent events with current exposure, but can also be used to estimate the probability of a hazard occurring at a location with a specific intensity (UNISDR, 2013).

Hazard maps showplace and level of hazard. The intensity of a hazard is usually determined based on data on, e.g., a hazard's frequency and magnitude of occurrence. These differ due to the specific characteristics of each hazard. *(Fleischhauer et al., 2005)*

Vulnerability maps are the result of mapping the various types of vulnerabilities. Each type of vulnerability is characterized by several indicators/variables. Vulnerability indicators are multi-dimensional, dynamic in time, scale-dependent and site-specific different indicators are selected in the various vulnerability assessments studies *(Hahn, Villgran De León, et al. 2003, Villgran De León, 2006, Lummen, Yamada, 2014)*. The choice of indicators to analyze also depends on data availability. Data involving socio-economic characteristics or their changes are often not available at local levels, or they are not available at all.

The final of risk analysis is to determine the level of risk; which may be determined by qualitative or quantitive methods. Qualitative risk assessment through a risk matrix is more common for risk mapping. The risk matrix combines three kinds of information *(Papathoma – Kohle et al., 2016)*:

- *information regarding the hazard (probability),*

- *information regarding the impact/vulnerability (consequences),*

- *information regarding the risk criteria of the end-user.*

There are two approaches to risk mapping in published studies. Both are based on a set of maps. A separate map with weight assignment is generated for each variability. However, the first approach clearly defines variables separately for hazard and individually for at least 1 type of vulnerability. There are not unusual studies focused only on 1 type of vulnerability, eg. to the environment, but also 2 or 3 types. A common technique is to assign proper weights to each variable concurring to hazard and vulnerabilities, combining those in a risk matrix for the result risk map. In such a case, the map is created based on the diagram of Figure 1.

The second approach is to define a group of variables without dividing them into hazard and vulnerability variables so that the integrated hazard and vulnerability maps are not created, but directly the risk map (for example in study Gheshlaghi, 2019). It should be noted that there are also more complex studies where the third parameter is included in the risk assessment, namely the capacity according to the mathematical formula number 3.

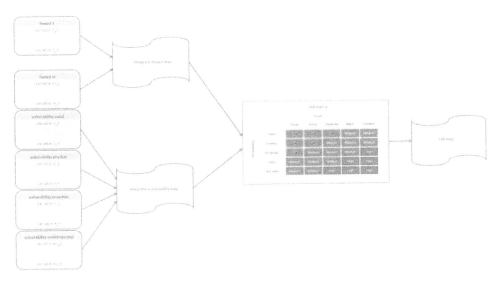

Figure 1. The process of risk map creating beset on hazard and vulnerability maps.

4 CONCLUSION

In the literature, risk mapping studies are quite common in various fields of interest. Risk maps are already published on city websites to inform and increase the safety of the population. Often, the content of the map, and the way it is processed is up to the author of map. It should be noted that the content of analyzes also depends on the availability of input data. Another problem is the use of the concept of danger and risk as synonyms. In this case, hazard maps are referred to as risk maps, which essentially does not take into account the amount of hazard impact. ISO 31 100 clearly defines terms related to risk analysis, but terms related to risk maps are not explicitly defined.

ACKNOWLEDGEMENT

This paper was written thanks to supporting from projects: SKHU/1601/4.1/187: "Logistic support system for flood crisis management in the Hernád/Hornád catchment" and from project APVV-18-0351: "Risk assessment of the parasitozoonoses occurrence using multicriteria analysis approaches".

REFERENCES

Albano, R., Sole, A. 2018. Geospatial Methods and Tools for Natural Risk Management and Communications. *International journal of geo-information*, 7 (12).

Aven, T. 2016. Risk assessment and risk management: Review of recent advances on their foundation. *European Journal of Operational Research*, 253(1): 1–13.

Birkmann, J. 2006. Measuring vulnerability to promote disaster-resilient societies: Conceptual frameworks and definitions. *In: Measuring vulnerability to natural hazards: Towards disaster resilient societies*, 1: 9–54.

Ciurean, R. L., Schröter, D., Glade, T. 2013. Conceptual frameworks of vulnerability assessments for natural disasters reduction. *Approaches to disaster management-Examining the implications of hazards, emergencies, and disasters*. IntechOpen.

Di Salvo, C., Pennica, F., Ciotoli, G., Cavinato, G. P. 2018. A GIS-based procedure for preliminary mapping of pluvial flood risk at metropolitan scale. *Environmental modeling & software*, 107: 64–84.

Fleischhauer, M., Greiving, S., Schlusemann, B., Schmidt-Thomé, P., Kallio, H., Tarvainen, T., Jarva, J. 2005. Multi-risk assessment of spatially relevant hazards in Europe. In *ESPON, ESMG symposium, Nürnberg*.

Gheshlaghi, H. A. (2019). Using GIS to Develop a Model for Forest Fire Risk Mapping. *Journal of the Indian Society of Remote Sensing*, 47(7),1173–1185.

Hahn, H., J. C. Villgran DeLeón, et al. 2003. Indicators and Other Instruments for Local Risk Management for Communities and Local Governments. *In: Local Risk Management for Communities and Local Governments*. G. f. I. The German Technical Cooperation Agency.

Hokstad, P., Steiro, T. 2006. Overall strategy for risk evaluation and priority setting of risk regulations. *Reliability Engineering & System Safety*, 91(1): 100–111.

Lummen, N. S., Yamada, F. (2014). Implementation of an integrated vulnerability and risk assessment model. *Natural hazards*, 73(2),1085–1117.

Papathoma-Köhle, M., Promper, C., Glade, T. 2016 -a. A common methodology for risk assessment and mapping of climate change related hazards—implications for climate change adaptation policies. *Climate*, 4(1), 8.

Papathoma-Koehle, M., Promper, C., Bojariu, R., Cica, R., Sik, A., Perge, K., Birsan, M. V. 2016- b. A common methodology for risk assessment and mapping for south-east Europe: an application for heat wave risk in Romania. *Natural Hazards*, 82(1), 89–109.

Sari, D. A. P., Innaqa, S., Safrilah, 2017. Hazard, vulnerability and capacity mapping for landslides risk analysis using geographic information system (GIS). *In: IOP conference series: materials science and engineering*. Vol. 209. No. 1. IOP Publishing.

Villgran de León JC. 2006. Vulnerability—a conceptual and methodological review. UNU EHS, no4/ 2006, Bonn, Germany.

UNISDR (International Strategy for Disaster Reduction). 2002. Living with risk. A global Review of disaster reduction initiatives. UN Publications, Geneva.

UNISDR (International Strategy for Disaster Reduction), 2004. *Living with Risk: A Global Review of Disaster Reduction Initiatives* Version. UN Publications, Geneva.

UNISDR (United Nations International Strategy for Disaster Reduction), 2013. *Global Assessment Report on Disaster Risk Reduction*. UN Publications, Geneva.

Advances and Trends in Geodesy, Cartography and Geoinformatics II –
Molčíková, Hurčíková, & Blišťan (eds)
© 2020 Taylor & Francis Group, London, ISBN 978-0-367-34651-5

Participatory methodology for development of geo-ontologies: Initial study

O. Čerba
University of West Bohemia, Plzeň, Czech Republic

ABSTRACT: Geo-ontologies (ontologies of geographical objects) are an essential tool from the perspective of the semantics of seemingly clear and unambiguous geographic data and information. They describe geographic features and type of features, and so improve the understanding of the structure of the geographic world among different subjects (cultures, languages, nations, social groups and others). This property is essential from the view of geographic data manipulation, sharing, changing and interconnecting.

Just the content sharing as the central principle of general ontologies and the focus on standard features in the specific case of geo-ontologies is a challenge to improve a methodology of geo-ontologies development by an implementation of the participatory approach. Moreover, several existing methodologies contain pieces of such process (for example, user-focused evaluation in On-To-Knowledge methodology or competency questions in Grüninger & Fox methodology).

This article describes the initial study of the development of the methodology for the construction of geo-ontologies. The methodology will be wholly based on a participatory approach. Therefore this article is focused on a review of existing methodologies for ontological engineering from the view of participatory steps.

1   INTRODUCTION

One of the specifics of geographical objects is their duality. On the one hand, objects such as river or mountain are understandable for everyday users, and each common language uses these terms frequently. On the other hand, these objects are related to very complicated scientific information. The article Smith&Mark (2001) describes this dual view in more detail.

Geo-ontologies (ontologies of geographical objects) are an essential tool from the perspective of the semantics of seemingly clear and unambiguous geographic data and information. They can describe geographic features or geographic objects by a formal and formalized way. Such description using an explicit semantics improves the understanding of particular structures of the geographic world. The comprehensibility is essential from the view of geographic data manipulation, sharing, changing and interconnecting.

The content sharing as the central principle of general ontologies and the focus on commonly known and seemingly non-expert features (such as mountains, regions or rivers) in the specific case of geo-ontologies is a challenge to improve a methodology of geo-ontologies development by an implementation of the participatory approaches (Slocum & Thomas-Slayter, 1995). Moreover, several existing methodologies contain pieces of such process (for example, competency questions).

The primary goal of this article is to describe a possible influence of participatory approach to methodologies related to developing knowledge bases and ontologies on the geographical domain. The discovering of points of intersection of ontological engineering and participatory approach is based on a detailed comparison of particular steps, methods, tools and techniques of existing methodologies for the development of ontologies from the perspective of participation of various interesting subjects (domain experts, ontological engineers, users or public).

The article is structured as follows. The research section introduces the essential terms and studies dealing with the methodologies of ontological engineering. The methodology shows the processes of identifying and evaluating traces connected to a participatory approach in existing methodologies. The results section presents outputs of this research (particular pieces of participatory processes in existing methodologies for the development of geo-ontologies and comparison of methodologies). These results are discussed in the following section.

2 RESEARCH

In 1997, Borst extended the Gruber's original definition of ontology (Gruber, 1993) and described an ontology as a "formal specification of a shared conceptualization" (Borst, 1997). According to (Guarino et al., 2009) "this definition additionally required that the conceptualization should express a shared view between several parties, a consensus rather than an individual view". This understanding of ontologies is significant from the perspective of implementation of participation and involvement of various experts, stakeholders and public.

Geo-ontology can be described as an ontology dealing with "the totality of geospatial concepts, categories, relations, and processes and with their interrelations at different resolutions" (Mark et al., 2004). This definition shows essential characteristics of geo-ontologies – features related to the earthbound space and overlap to many other domains of human activities. The other specifics of geo-ontologies are mentioned in An, Y., & Zhao (2007), Henriksson et al. (2008) or Janowicz (2012).

Because geo-ontologies are the subset of general ontologies, and there are not specific and intricate methodologies for the construction of geo-ontologies, the research of relevant methodologies is focused on standard ontologies. During the last thirty years, many various methodologies for the development of ontologies emerged. Based on the in-depth research of studies (Jones et al., 1998, Fernández-López, 1999, Corcho et al., 2003, Cristani & Cuel, 2005, Iqbal et al., 2013) focused on description and comparison of existing methodologies of ontological engineering, the following group of methodologies was chosen. The final set contains eight methodologies (TOVE, Enterprise Model Approach, METHONTOLOGY, Methodology published by Bernaras et al., On-To-Knowledge Methodology, DOLCE, Ontology Development 101 and UPON), which are mentioned in the majority of researched studies or are essential from the development of ontologies. Other methodologies (for example Cyc) are not described in the result section, but they are considered and mentioned in the discussion.

3 METHODOLOGY

The goal of this article is to describe a possible influence of participatory approach to methodologies related to developing knowledge bases and ontologies on the geographical domain. The process of the research is composed of the following three steps: (1) The research of methods focused on the development of ontologies and geo-ontologies; (2) Identification and description of participatory steps, techniques or tools in particular methodologies; (3) Evaluation of particular methodologies from the view of the participation of various subjects in the process of development.

The articles (Fernández-López, 1999, Corcho et al., 2003, Cristani & Cuel, 2005, Iqbal et al., 2013) use a similar approach for comparison on methods of ontological engineering. The following section describes the results of the implementation of this procedure on selected methodologies used in geo-ontological engineering.

4 RESULTS

Results section has the following structure. The particular methodologies for the development of ontologies are shortly described from the perspective of participatory methods. After that, participatory activities related to ontology construction are derived.

TOVE (Toronto Virtual Enterprise; in publication Fernández-López, 1999 named as Methodology By Grüninger And Fox) (Grüninger & Fox, 1995a, Uschold & Grüninger, 1996, Grüninger & Fox, 1996) is composed of six necessary steps – motivating scenarios, informal competency questions, terminology specification, formal competency questions, axiom specification and completeness theorems. From the view of the involvement of various subjects, steps can be divided into three groups:

1. Expert activities (formal competency questions, axiom specification),
2. steps, which can partially use external involvement for example for searching alternatives or evaluation (specification of terminology, especially the sub-step Getting informal terminology and completeness theorems),
3. processes that are strongly related to the participation of a broad spectrum of stakeholders (motivating scenarios, informal competency questions).

Enterprise Model Approach (Uschold & King, 1995, Uschold, 1996) (in publications Fernández-López, 1999 and Corcho et al. 2003 named as Enterprise Ontology) is straightforward four-steps methodology, which can be supported by the participation of external subjects in all steps (except step 3 – formalisation, which consists in the construction of definitions and logical axioms). Various subjects dealing with developed geo-ontology can contribute effectively to the identification of purpose and scope of ontology (steps 1 and 2) as well as to formal evaluation (step 4).

METHONTOLOGY (Gómez-Pérez et al., 1996, Fernández-López et al., 1997) is similarly to TOVE focused not only on development but also on maintenance. This methodology includes seven steps (specification, knowledge acquisition, conceptualisation, integration, implementation, evaluation and documentation). Many of these steps are connected with expert work, but at least specification and partially knowledge acquisition and evaluation represent phases for implementation of participatory techniques.

Methodology published by Bernaras, Laresgoiti and Corera (Bernaras et al., 1996) was implemented in the KACTUS project. Therefore it is described as CommonKADS and KACTUS or KACTUS. The methodology is straightforward and similar to the Enterprise Model Approach. It contains three very general steps – Specification of the application (this phase is crucial from the perspective of participatory approach), Preliminary design based on relevant top-level ontological categories and Ontology refinement and structuring.

Similarly to previous methodologies, On-To-Knowledge Methodology (OTK Methodology) (Staab et al., 2001, Sure et al., 2004) is represented by a linear process, which is composed of several steps – Feasibility study, Ontology kickoff, Refinement, Evaluation and Maintenace. The particular steps are connected with the involvement of a broad spectrum of external subjects. For example "Concept elicitation with domain experts" (Staab et al., 2001) is crucial for the Refinement phase or "Requirement specification" (Staab et al., 2001) is an integral part of Ontology kickoff.

The methodology used for the development of DOLCE (Descriptive Ontology for Linguistic and Cognitive Engineering) (Masolo et al., 2002) "is based on a few starting points for building new ontologies" (Cristani & Cuel, 2005). Above all, the first point (determination of features from the domain, which will be modelled) is closely connected to participation approach, because it requires involvement domain experts.

Ontology Development 101 (Noy & McGuinness, 2001) is designed primarily for developers. It is focused on concrete technical steps. The methodology helps to construct ontologies in environments such as Protégé, Ontoligua or Chimaera (Cristani & Cuel, 2005). Adequate to previous methodologies, also Ontology Development 101 starts with a determination of the domain and scope of the ontology. This step requires involvement not only ontology engineers, domain experts, but above all users and stakeholders. This initial phase is connected with competency questions mentioned in (Grüninger & Fox, 1995b). Ontology Development 101 contains other activities related to participatory techniques and methods (for example enumeration of essential terms in the ontology).

UPON methodology (De Nicola et al., 2005) "takes advantage of the Unified Process (UP) and adopts the Unified Modeling Language (UML) as well" (Cristani & Cuel, 2005). This methodology is composed of five workflows (Requirements, Analysis, Design, Implementation and Test), which are realised in iterative cycles containing four phases (Inception, Elaboration, Construction and Transition). Figure 1 in (De Nicola et al., 2005) shows the involvement of domain experts in particular workflows. The level of involvement is high in first two workflows, then it goes down, but at the last workflow (Test) the importance of participation of domain experts increases.

5   DISCUSSION

From previous paragraphs, it is evident that the participatory approach is fitting for these parts of the development of geo-ontologies, which are not related to coding, formalisation, data models building or populations of ontologies. The following steps of ontological development corresponding with methodologies mentioned above represent ideal fields for interconnection of participatory methods and ontological engineering:

*   The first phase of designing of geo-ontologies, for example, competency questions (described in (Grüninger & Fox, 1995b), examples in (Noy & McGuinness, 2001) or tasks related to a delimiting of scopes or purposes of ontologies.
*   Acquiring inputs (classes, instances and properties) to ontologies, for example, the second part of the process used to build the Cyc Knowledge Base (Lenat & Guha 1989) is based on knowledge acquisition performed by humans.
*   Evaluation, testing and other activities focused on acquiring feedback and information from stakeholders and experts from associated domains.

Taking into consideration the specifics of geo-ontologies, it is necessary to realise that they are very appropriate for the implementation of participatory methods. There are two main reasons:

*   Features of geo-ontologies are a part of the universal human understanding of the world.
*   Geo-ontologies cross and interconnect isolated domains not only from geography but also from other disciplines.

For purposes of geo-ontological development, various participatory techniques such as brainstorming, structured, non-structured or informal interviews (these techniques are mentioned in (Fernandez et al., 1997) explicitly) or group sketching should be incorporated to a process of building ontologies. This research focused on new or improvement methodology will follow-up this initial study.

6   CONCLUSIONS

This article is the initial study for the development of an innovative methodology for the construction of ontologies on the geographic domain (geo-ontologies). Because of specifics of geo-ontologies (multi-disciplinarity, using familiar concepts and interconnection to many human activities), such methodology has to take into consideration not only work of ontological engineers but also other interesting subjects such domain experts, stakeholders, users and public as well. Therefore the implementation of participatory methods, techniques and tools is a proper way of how to improve existing methodologies. The main goal of the article is to provide arguments for the close interconnection of geo-ontologies and participatory methods supporting their development. These arguments can complete and accelerate the theory of the construction of geo-ontologies.

The research published in this article provides a unique comparison of parts of the existing methodologies of ontologies engineering, which are related to the participatory approach. It shows that the main steps connected to the involvement of external subject deal with designing of geo-ontologies, acquiring input data, information and knowledge as well as evaluation, testing and providing feedback. From the article, it is evident that interconnection of methodologies dealing participatory approach and development of ontologies is an ideal touching point for meetings of ontological engineers, domain experts, experts from other geo-disciplines, other stakeholders, users and public.

ACKNOWLEDGMENTS

This publication was supported by the project LO1506 of the Czech Ministry of Education, Youth and Sports.

REFERENCES

An, Y., & Zhao, B. (2007). Geo ontology design and comparison in geographic information integration. In *Fourth International Conference on Fuzzy Systems and Knowledge Discovery (FSKD 2007)*, Vol. 4, pp. 608–612. IEEE.

Bernaras, A. et al. 1996. Building and reusing ontologies for electrical network applications. *Proc. of ECAI 96*, 1996, 298–302.

Borst, W. N. 1997. *Construction of engineering ontologies for knowledge sharing and reuse.*

Corcho, O., Fernández-López, M., Gómez-Pérez, A. 2003. Methodologies, tools and languages for building ontologies. where is their meeting point? *Data & knowledge engineering.* 46:41–64.

Cristani, M., Cuel, R. 2005. A survey on ontology creation methodologies. *International Journal on Semantic Web and Information Systems (IJSWIS).* 1:49–69.

De Nicola, A., Missikoff, M., & Navigli, R. 2005. A proposal for a unified process for ontology building: UPON. In *International Conference on Database and Expert Systems Applications.* p. 655–664. Springer, Berlin, Heidelberg.

Fernández-López, M. 1999. Overview of methodologies for building ontologies. In *Proceedings of the IJCAI99 Workshop on Ontologies and Problem Solving Methods Lessons Learned and Future Trends.* 13 p.

Fernández-López, M., Gómez-Pérez, A., & Juristo, N. 1997. Methontology: from ontological art towards ontological engineering. In www.aaai.org – *AAAI Technical Report SS-97-06. 1997, AAAI-97 Spring Symposium Series, Proceedings of the Ontological Engineering AAAI-97 Spring Symposium Series*, 8 p.

Gómez-Pérez, A., Fernández, M., & Vicente, A. D. 1996. Towards a method to conceptualize domain ontologies. In *12th European Conference on Artificial Intelligence (ECAI'96), Proceedings Workshop: Ontological Engineering*, 11 p.

Gruber, T. R. 1993. A translation approach to portable ontology specifications. *Knowledge acquisition* 5:199–220.

Grüninger, M., & Fox, M. S. 1995a. Methodology for the design and evaluation of ontologies. In *workshop on Basic Ontological Issues in Knowledge Sharing*, IJCAI-95, Montreal, 10 p.

Grüninger, M., & Fox, M. S. 1995b. The role of competency questions in enterprise engineering. In *Benchmarking—Theory and practice.* p. 22–31. Springer, Boston, MA.

Gruninger, M., & Fox, M. S. 1996. The logic of enterprise modelling. In *Modelling and Methodologies for Enterprise Integration.* p. 140–157. Springer, Boston, MA.

Guarino, N., Oberle, D., Staab, S. 2009. What is an ontology? In: *Handbook on ontologies.* Springer, p 1–17.

Henriksson, R., Kauppinen, T., & Hyvönen, E. (2008). Core geographical concepts: Case Finnish geo-ontology. In *Proceedings of the first international workshop on Location and the web* (pp. 57–60). ACM.

Iqbal, R., Murad, M. A. A., Mustapha, A., Sharef, N. M. 2013. An analysis of ontology engineering methodologies: A literature review. *Research journal of applied sciences, engineering and technology.* 6:2993–3000.

Janowicz, K. (2012). Observation-driven geo-ontology engineering. *Transactions in GIS*, 16(3), 351–374.

Jones, D., Bench-Capon, T., Visser, P. 1998. *Methodologies for ontology development.*

Lenat, D. B., & Guha, R. V. 1989. *Building large knowledge-based systems; representation and inference in the Cyc project.*

Mark, D. M., Smith, B., Egenhofer, M., Hirtle, S. 2004. Ontological foundations for geographic information science. *Research Challenges in Geographic Information Science.* 335–350.

Masolo, C., Borgo, S., Gangemi, A., Guarino, N., & Oltramari, A. 2002. Wonderweb deliverable d17. *Computer Science Preprint Archive*, 2002(11),74–110.

Noy, N. F., & McGuinness, D. L. 2001. Ontology development 101: A guide to creating your first ontology. In *Stanford Knowledge Systems Laboratory, Technical Report KSL-01-05 and Stanford Medical Informatics Technical Report SMI-2001-0880.* 25 p.

Slocum, R., & Thomas-Slayter, B. (1995). Participation, empowerment and sustainable development; A brief history of participatory methodologies. In *Power, process and participation: Tools for change* (pp. 1–16). Practical Action Publishing.

Smith, B., & Mark, D. M. 2001. Geographical categories: an ontological investigation. *International journal of geographical information science*, 15(7), 591–612.

Staab, S., Studer, R., Schnurr, H. P., & Sure, Y. 2001. Knowledge processes and ontologies. *IEEE Intelligent systems*, 16(1), 26–34.

Sure, Y., Staab, S., & Studer, R. 2004. On-to-knowledge methodology (OTKM). In *Handbook on ontologies*. (p. 117–132. Springer, Berlin, Heidelberg.

Uschold, M. 1996. Building ontologies: Towards a uni ed methodology. In *Proceedings of 16th Annual Conference of the British Computer Society Specialists Group on Expert Systems.*

Uschold, M., & Gruninger, M. 1996. Ontologies: Principles, methods and applications. *The knowledge engineering review*, 11(2), 93–136.

Uschold, M., & King, M. 1995. *Towards a methodology for building ontologies.* p. 19. Edinburgh: Artificial Intelligence Applications Institute, University of Edinburgh.

Advances and Trends in Geodesy, Cartography and Geoinformatics II –
Molčíková, Hurčíková, & Blišťan (eds)
© 2020 Taylor & Francis Group, London, ISBN 978-0-367-34651-5

Determination of the effect of solar radiation in spatial predictive modelling of the bark beetle occurrence using several mathematical methods

R. Ďuračiová & V. Jakócsová

Department of Theoretical Geodesy, Faculty of Civil Engineering, Slovak University of Technology, Bratislava, Slovak Republic

ABSTRACT: Spatial predictive models are usually based on a combination of multiple input factors, using various methods for spatial predictive modelling. Some of them automatically calculate the weights of individual input factors, which also determine their effect on the resulting model, others require their explicit input. The paper deals with the comparison of several methods of weights determination in spatial predictive modelling. The case study focuses mainly on the solar radiation parameter and its effect on forest damage caused by bark beetle attacks in the High Tatra Mountains. The aim of the paper is the determination of weighting factor of potential global solar radiation in spatial predictive modelling of susceptibility to the occurrence of bark beetle spots using various mathematical methods. The advantage of applying several methods to determine the weights of the input factors is also the possibility of checking the accuracy of the calculation. We recommend using Student's t-test and then calculating the Gini indices or deriving the fuzzy membership function for each input factor based on frequency histograms. The result is also confirmation and quantification of the effect of global solar radiation on the occurrence of bark beetle spots in the High Tatra Mountains.

1 INTRODUCTION

The aim of spatial predictive modelling in geographic information systems (GIS) is to predict the future behavior or occurrence of a phenomenon in unknown areas based on its occurrence in known areas (Gomarasca 2009). Spatial predictive models are based on a combination of multiple input factors, such as slope, aspect, elevation, proximity to some objects, etc. The paper deals with the comparison of some methods for the determination of the weights of the input factors in spatial predictive modelling. The case study focuses mainly on the solar radiation parameter and its effect on forest damage caused by bark beetle attacks in the High Tatra Mountains. The effect of solar radiation on bark beetle attacks in forests was supposed or described in some papers (Mezei et al. 2012), (Öhrn 2012), (Chen et al. 2015), (Jeníček et al. 2017), (Jakuš et al., 2017). For example, Kissiyar et al. (2005) and Jakuš et al. (2005) applied solar radiation as one of site characteristics in computation of probability of bark beetle attack in the High Tatra Mountains using multiple linear regression. Mezei et al. (2019) compared some meteorological parameters and potential solar radiation in relation to their impact on the occurrence of bark beetle spots. The used parameters were mutually correlated, therefore they applied univariate logistic regression for each parameter separately to analyze their individual relationship with the occurrence of bark beetles. The aim of this paper is to confirm this impact and to quantify it by applying several mathematical methods, as well as to compare their processing, results and objectivity of the results.

2 METHODS

To achieve the aim of this work we calculated potential global solar radiation based on digital elevation model (DEM) using *Area Solar Radiation tool* in software ArcGIS (Esri, Redlands, CA, USA). Global solar radiation was calculated as a yearly sum with a 14 day interval.

For determining the effect of potential global solar radiation as the input factor in the spatial predictive modelling, we proposed application of several mathematical methods, namely a statistical hypothesis test, bivariate and multivariate analyses, a statistical measure of distribution, and the fuzzy set theory. The specific methods are briefly described below.

The Student's t-test is a parametric statistical hypothesis test that assesses whether two variable's means (i.e., means from independent groups) are statistically different from each other (Snedecor and Cochran 1989). In our case, it is useful for testing the variation of the phenomenon distribution (potential global solar radiation as an independent variable) in areas where the predicted phenomenon (the occurrence of bark beetle spots as a dependent variable) compared to the study area.

Bivariate and multivariate analyses are well known quantitative statistical methods to investigate relationships between data samples. Bivariate analysis (Babbie 2010) is concerned with the relationships between pairs of variables such as the relationship of the dependent versus independent variables in regression. In our case, bivariate analysis is employed to examine and quantify the relationships between the dependent variable (value of a predictive model in GIS) and each of the independent variables, e.g. potential global solar radiation.

Logistic regression is a special form of multivariate regression analysis, in which the dependent variable is a nonmetric, dichotomous (binary) variable (Hair et al. 2010). It is also the case of the predictive models in GIS, where we predict the occurrence of phenomena, i.e., whether or not it is present in a given site (raster cell in the GIS software environment). Note that more known linear regression is not suitable for this type of classification problem.

The Gini coefficient, sometimes called *Gini index*, or *Gini ratio* (Gini 1912), is a measure of statistical dispersion used primarily in economics, but it is also useful in geostatistics (Wackernagel 2003). We applied it to calculate the weights of input factors in spatial predictive modeling (for each parameter separately) (Lieskovský et al. 2015).

The fuzzy set theory (Zadeh 1965) is primarily intended for modeling uncertainty, so it is often used in multi-criteria decision making, including predictive modeling in GIS. When applying the fuzzy sets theory in spatial predictive modelling in our study, we created fuzzy sets based on frequency histograms. Fuzzy sets in multi-criteria decision making can also be determined using expert judgment, but this is a more subjective way.

3 CASE STUDY

In the experiment we applied all above mentioned methods for determining the weighting factors, respectively the effect of each input parameter (e.g., solar radiation) on the occurrence of modeled phenomenon (in this case bark beetle attack). Therefore, we can reveal and quantify the relationship between the calculated potential global solar radiation and bark beetle spots in the High Tatra Mountains (Jakuš et al. 2005), (Havašová et al. 2015).

3.1 *Data*

The case study area is the western part of the High Tatra Mountains (Figure 1). We calculated potential global solar radiation based on the Shuttle Radar Topography Mission (SRTM) DEM of 30 m resolution (http://dwtkns.com/srtm30m). As other factors in predictive modelling, the normalized difference vegetation index (NDVI) derived from Landsat satellite data (30 m resolution) and distance to old spots we applied (see (Havašová et al. 2015) for details and data availability). For simplicity and clarity of the example, we used only three input factors, but in fact they can be used much more in predictive modelling of forest damage

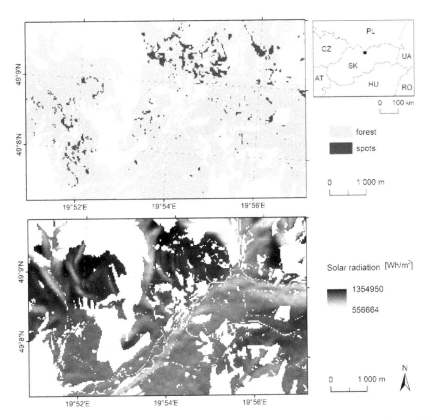

Figure 1.   Detail of the map of bark beetle spots (upper) in the High Tatra Mountains from 2009, data source: (Havašová et al. 2015) and detail of the map of potential global solar radiation calculated in ArcGIS 10.5 using *Area Solar Radiation tool* (bottom).

(Kissiyar et al. 2005), (Jakuš et al. 2017). We used data from a period of 3 years after calamity in the High Tatra Mountains (2004). For prediction, data from 2008 we used. Data from 2007 were used only to calculation of distances from old bark beetle spots to new ones (Jakuš et al. 2017). Validation of prediction was based on known bark beetle spots from 2009 (Havašová et al. 2015). The sample of bark beetle spots from 2009 is shown on the upper map in Figure 1.

3.2   *Data processing*

In the case study, we used three data layers related to 2008 (potential solar radiation (S), distance from old spots (D), and NDVI (N)) to determine their impact on the predictive model, but in statistical analyses we focused on the potential global solar radiation parameter.

   We realized the main data preparation and processing in the ArcGIS 10.5 software environment. The calculation of potential global solar radiation was performed on 414536 cells representing healthy forest and 4000 cells representing bark beetle spots (Jakócsová 2018). The sample of the map of calculated potential global solar radiation is shown in Figure 1 (bottom).

   For statistical data analyses, software Microsoft Excel and the Python programming language we used. T-test, logistic regression, bivariate analysis, and Gini index were calculated by the r.learn.ml module (using the Python scikit-learn package). Analysis by the fuzzy set theory was processed in ArcGIS 10.5 using tools *Fuzzy Overlay* and *Fuzzy Membership*.

4 RESULTS

According to Student's *t*-test, the solar radiation parameter in 2008 (Figure 2) has significantly different values in the forest and in the spruce bark beetles (*p*-value = 0) (Jakócsová 2018). Based on this we quantified the significance of this parameter using following analyses.

The coefficients of reclassified parameters in bivariate analysis led to the weight of the solar radiation parameter $w_{BS} = 0.11$. The weight of the distance parameter and the weight of the NDVI parameter were $w_{BD} = 0.56$ and $w_{BN} = 0.34$, respectively. The number of reclassification categories for these two parameters was 50.

Logistic regression results in regression coefficients (-0.3805, -1.1216, 0.0113). Their interpretation is more difficult compared to other statistical methods used. Therefore, we applied this method to the layers reclassified into 10 categories of suitability, although it is a partially subjective step. On the basis of these we determined the weights of all parameters: $w_{RS} = 0.2$, $w_{RD} = 0.5$, $w_{RN} = 0.3$.

Using the Gini index (Figure 3), we determined the weight w_{GS} of the solar radiation parameter at 0.13 (when applying three input factors: solar radiation (Gini index 0.23, $w_{GS} = 0.13$), distance from old spots (Gini index 0.95, $w_{GD} = 0.54$), and NDVI (Gini index 0.57, $w_{GN} = 0.33$).

In case of application of the fuzzy set theory, we used the relative frequency histograms as the basis for derivation of membership function. The histograms of the frequency of values of potential global solar radiation in the healthy forest versus bark beetle spots are shown in Figure 4. On their basis (specifically the *Spots* function shape) we used the Fuzzy Large membership function:

$$\mu(x) = \frac{1}{1 + \left(\frac{1}{f2}\right)^{f1}},$$ (1)

where $\mu(x)$ is a degree of membership of value x, $f2$ is the midpoint, and $f1$ is the spread of the function. The midpoint defines the crossover point (membership of 0.5). Based on functions shown in Figure 4, we derived parameters $f2 = 1\ 150\ 000$ and $f1 = 5$. The value of $f2$ was derived from the intersection of the functions *Spots* and *Forest* in Figure 4 and the value of $f1$ was derived from the intersection of the function *Spots* and the x-axis. The effect of the solar radiation parameter on the resulting predictive model is mainly expressed by $f1$. This value can then be used to calculate the weight, but this calculation requires analysis of all input factors.

The results of analyses are summarized in Table 1. The Table 1 also provides information on which methods require subjective intervention in the form of expert knowledge (e.g. the number of categories in case of bivariate analysis) as well as which of them are suitable for calculating the impact of each parameter separately.

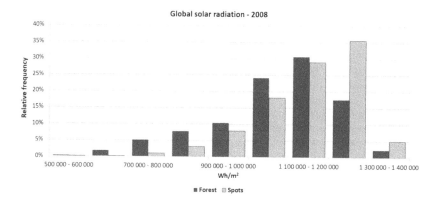

Figure 2. Relative frequency histograms of values of potential global solar radiation on the healthy forest (*Forest*) and the bark beetle spots (*Spots*).

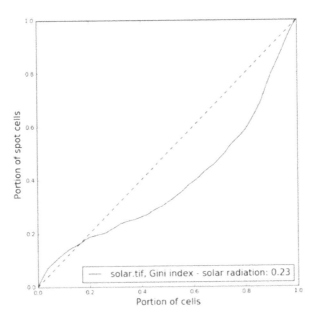

Figure 3. The Gini index of the potential global solar radiation parameter in predictive modelling of forest damage.

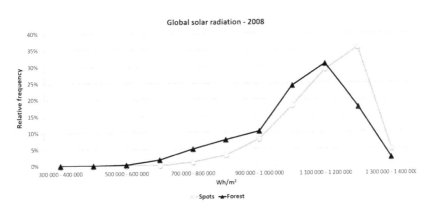

Figure 4. Relative frequency of potential global solar radiation as a basis for the formation of the membership function for the fuzzy set.

Table 1. Comparison of methods for determining the effect of the input factors on the predictive model.

Methods	Need of expert knowledges	Possibility to process each parameter separately	Results
t-test	no	yes	$p = 0$
Bivariate analysis	yes	yes	$w_{BS} = 0.11$
Logistic regression	yes	no	$w_{RS} = 0.10$
Gini index	no	yes	$w_{GS} = 0.13$, $G = 0.23$
Fuzzy set theory	yes	yes	*Large: $f1 = 5, f1 = 1\ 150\ 000$*

5 CONCLUSIONS AND DISCUSSION

Based on our analysis, we can confirm in accordance with (Mezei et al. 2019) the importance of modeling of solar radiation in the GIS environment as a suitable parameter to predict forest damage. Its influence on the resulting prediction model was confirmed by all methods used. Although the effect of solar radiation on the bark beetle spread was assumed and confirmed in previous papers (Kissiyar et al. 2005), (Chen et al. 2015), (Jakuš et al. 2017), (Mezei et al. 2019), in our work we designed and tested methods for its quantification. The aim of this paper was to calculate this effect on the predictive model of susceptibility to the occurrence of bark beetles, but it is important to note that the applied methods can also be applied to all other input factors (or independent variables) in predictive modelling in GIS. From the point of view of ease of processing, we recommend using Student's t-test and then calculating the Gini indices or deriving the fuzzy membership function for each input factor. Other methods require a comprehensive analysis of all input parameters together. The advantage of applying several methods to determine the weights of the input factors is also the possibility of checking the accuracy of the calculation. Note that the above results refer only to the analysis carried out in a small area and to a specific example of solar radiation analysis within one year. Complete analysis requires data processing over a longer period of time, in a larger area, and applying other factors affecting the spread of spruce bark beetle, such as age of forest, stand density, volume per hectare and so on (Jakuš et al. 2005). Moreover, in order to determine the impact of solar radiation more precisely, it is necessary to determine more precisely solar radiation, because the main limitation of our work was the resolution of the DEM (30 m) used for its calculation.

ACKNOWLEDGEMENT

This work was supported by Grant No. 1/0300/19 of the VEGA Grant Agency of the Slovak Republic.

REFERENCES

Babbie, E.R. 2010. The practice of social research. 12th Edition. Belmont: Wadsworth Publishing.
Chen, F. Zhang, G., Barlage, M., Zhang, Y., Hicke, J. A., Meddens, A., Zhou, G., Massman, W. J. & Frank, J. 2015. An Observational and Modeling Study of Impacts of Bark Beetle–Caused Tree Mortality on Surface Energy and Hydrological Cycles. *Journal of Hydrometeorology* 16: 744–761.
Gini, C. 1912. Variabilità e mutabilità. Reprinted in Pizetti, E.; Salvemini, T. (ed.), 1955. Memorie di metodologica statistica. Rome: Libreria Eredi Virgilio Veschi.
Gomarasca, M.A., 2009. Basics of Geomatics., New York: Springer-Verlag.
Hair, J. F., Black, W. C., Babin, B. J. & Anderson, R. E. 2010. Multivariate data analysis. 7th ed. New York: Macmillan Publishing Company.
Havašová, M., Bucha, T., Ferenčík, J. & Jakuš, R. 2015. Applicability of a vegetation indices-based method to map bark beetle outbreaks in the High Tatra Mountains. *Annals of Forest Research* 58: 295–310.
Havašová, M., Ferenčík, J. & Jakuš, R. 2017. Interactions between windthrow, bark beetles and forest management in the Tatra national parks. *Forest and Ecology Management* 391: 349–361.
Hebák, P., Hustopecký, J., Jarošová, E. & Pecáková, I. 2005. Vícerozměrné statistické metody (1), (2), (3). Praha: Informatorium.
Jakócsová, V. 2019. Výpočet solárnej radiácie pomocou aplikácie a modifikácie softvérových nástrojov GIS. Diplomová práca. Bratislava: Slovak University of Technology in Bratislava.
Jakuš, R., Ježík, M. & Blaženec, M. 2005. Prognosis of bark beetle attacks in TANABBO model. In Wojciech Grodzki (ed.), *GIS and databases in the forest protection in Central Europe*. Warsaw: Centre of Exellence PROFEST at the Forest Research Institute.
Jakuš, R., Blaženec, M., Koreň, M., Barka, I., Lukášová, K., Lubojacký, J. & Holuša J. 2017. TANABBO II model pro hodnocení rizika napadení lesních porostů lýkožroutem smrkovým Ips

typographus (L.) [Coleoptera: Curculionidae]. Lesnický průvodce 1/2017. Jíloviště: Výzkumný ústav lesního hospodářství a myslivosti, v. v. i.

Jeníček, M., Hotový, O. & Matějka, O. 2017. Snow accumulation and ablation in different canopy structures at a plot scale: using degree-day approach and measured shortwave radiation. *Acta Universitatis Carolinae Geographica* 52(1): 61–72.

Kissiyar, O., Blaženec, M., Jakuš, R., Willekens, A., Ježík, M., Baláž, P., Valckenborg, J.V., Celer, S. & Fleischer, P. 2005. TANABBO model: a remote sensing based early warning system for forest decline and bark beetle outbreaks in Tatra Mts. - overview. In Wojciech Grodzki (ed.), *GIS and databases in the forest protection in Central Europe.* Warsaw: Centre of Exellence PROFEST at the Forest Research Institute.

Lieskovský, T., Faixová Chalachanová, J., Ďuračiová, R., Blažová, E. & Karell, L. 2015. Archeologické predikčné modelovanie z pohľadu geoinformatiky. Metódy a princípy. Bratislava: Slovenská technická univerzita v Bratislave.

Mezei. P, Jakuš, R., Blaženec, M., Belánová, S. & Šmídt, J. 2012. The relationship between potential solar radiation and spruce bark beetle catches in pheromone traps. *Annals of Forest Research* 55(2): 243–253.

Mezei, P., Potterf, M., Škvarenina, J., Rasmussen, J. G. & Jakuš, R. 2019 Potential Solar Radiation as a Driver for Bark Beetle Infestation on a Landscape Scale. *Forests* 10, 604: 1–12.

Öhrn, P. 2012. The spruce bark beetle Ips typographus in a changing climate – Effects of weather conditions on the biology of Ips typographus. Introductory Research Essay No 18, Department of Ecology, SLU. Uppsala. 27 pp.

Snedecor, George W. and Cochran, William G. 1989. Statistical Methods, Eighth Edition, Iowa State University Press. 491 pp.

Wackernagel, H. 2003. Multivariate Geostatistics. An Introduction with Applications. Berlin Heidelberg New York: Springer-Verlag.

Zadeh, L. 1965. Fuzzy sets. *Information and control* 8: 338–353.

183

Advances and Trends in Geodesy, Cartography and Geoinformatics II –
Molčíková, Hurčíková, & Blišťan (eds)
© 2020 Taylor & Francis Group, London, ISBN 978-0-367-34651-5

Analysis of changes in map content rendering using vector tiles in web map application

R. Feciskanin & J. Fábry

Department of Cartography, Geoinformatics and Remote Sensing, Faculty of Natural sciences, Comenius University in Bratislava, Bratislava, Slovakia

ABSTRACT: In recent years, the publishing of geodata on the Internet has been a clear move towards using vector tiles. The technological development and the increase in the performance of client devices has allowed this significant change for web cartography A change in the form of published data is obvious. More importantly is the change in data processing the role of the web map client has become more significant. The difficulty of tasks for rendering vector data, where cartographic representation is provided from the side of the client, is greater than in raster data. The full topographic content of the map, which has traditionally been provided in the form of raster tiles, is also published in vector form. This means higher requirements for data preparation and for cartographic processing of vector data. Much attention is paid to the preparation of vector tiles and their optimization. Little interest is given to the negative effect of using vector tiles, such as higher load and time to render content on the web client. Although this effect is highly variable depending on the technology used, web browser capabilities and performance, we focused on quantitative evaluation and a comparison of the rendering process using vector and raster tiles. We followed the process of rendering tiles and creating the resultant image. We recorded the duration of each run of the rendering of gradually acquired tiles from the map server. We used OpenLayers library rendering events to retrieve the times. As tiles are requested parallel in groups and after obtaining the tiles follow their render (together with re-rendering of previously rendered tiles), we also recorded the number of transferred tiles, tile size and the time of transfer of the tiles from the server to the client. Average values of monitored variables from multiple measurements confirm differences in the rendering of raster and vector tiles and the importance of the influence of the change in rendering when changing tile type.

1   INTRODUCTION

The history of vector tiles in GIS applications dates back to the 1960s, but has seen a significant boom in its current form of distribution to client applications since 2010. In that year, they were used in Google Maps for Android (Google Inc., 2010). Before this, vector tiles appeared as designs for application, for example, in the form of professional articles (Antoniou et al., 2009) and submitted patents (Ellren & Gustafsson, 2010; Shi & Wu, 2010; Howell et al., 2011).

There is no standard for the distribution of spatial data in the the form of vector tiles, however, it is possible to use standard WMTS or WFS in order to create them (OGC, 2018a). The Open Geospatial Consortium (OGC) have coordinated several tests and are planning to include vector tile technology in the specifications of WMTS, WFS 3.0 and GeoPackage 1.2 (OGC, 2018b). The company Mapbox has contributed to the expansion of using vector tiles by creating an open specification called Mapbox Vector Tile (MVT) (Mapbox, 2017). This includes format support in ESRI products (Environmental Systems Research Institute, Inc. 2015), and has contributed to this specification more extensively. For convenience, we refer to the MVT format vector tile format as MVT.

The advantages of using vector tiles, which have contributed to the widespread use of web map applications such as graphic output quality, dynamic descriptions, individualization and change of cartographic symbolism, are often presented. However, they put increased demands on the client (web application and browser) – data processing, preparation of cartographic symbolism and more demanding rendering. In order to reduce complexity, a form of transmitted data can help reduce the burden on the client when preparing geometric rendering. This is done, for example, by defining geometric elements that are relative to the rendering cursor (methods *MoveTo*, *LineTo*) in the MVT specifications (Mapbox, 2017).

A complexity of actions has grown in the issue of using vector tiles and has shifted to the client. This is reminiscent of the work (Eriksson, Rydkvist, 2015), that seeks solutions to the relationship of data generalization in vector tiles and rendering speed. The work (Li et al. 2016) deals with the problem of rendering symbols on the edges between tiles. This is a concern for several university final papers that address the issue and are attempting to solve the impact of vector tile format on load time (Shang, 2015; Adamec, 2016).

The authors of this paper are attempting to contribute to the deepening of this issue, and therefore, the aim of the paper is to quantitatively express the changes in client activities brought about by the application of vector tiles in comparison to raster tiles. We focused on checking the time differences in rendering, data acquisition, and their size.

2   MATERIAL AND METHODS

In our experiment, we primarily monitored the course of data rendering and the time of transmission from the server to the client in vector and raster tiles using WMTS map service.

2.1   *Software*

We conducted the experiment in open source software QGIS 3.6, the map server GeoServer 2.14.2 and we created a client web application for independent measurement using Openlayers 5.3.0. The reason for choosing the GeoServer map server was the usability of the map service according to OGC standards with the possibility of publishing data in MVT format. In the case of the OpenLayers client library, support for both raster and vector tiles was important, as well as library events that helped us obtain the necessary layer data transfer from the server and the methods for rendering layers. The main tool selection criterion was the ability to use the same vector and raster tile tools in order to directly compare both types of tiles. An important fact was also the possibility of using a uniform record of cartographic symbolism (Mapbox style JSON version 8 created in Maputnik environment) for generating raster tiles on the server, as well as for drawing in the client application (with *ol-mapbox-style library*). The presented architecture solution is not the most sophisticated from the perspective of using vector tiles (eg. In comparison to instruments from Mapbox), but this would not enable an equivalent comparison to raster tiles. At the same time, the chosen solution highlighted and made it possible to present potential problems with the use of vector tiles. An additional tool was a simple parser to display vector tile content created using *pbf* and *vector-tile* libraries.

2.2   *Data and formats*

For our experiments, we used freely-available data from OpenStreetMap. The testing areas were the municipal districts of Bratislava and Trnava. Layers with a variety of types of geometry were used - point, line and polygon. Point layers include cities and communities for displaying labels. Line layers include roads and rivers. Polygon layers include buildings, land utilization, areal representation of waters (rivers, water areas) and areas of testing grounds.

From the data, a single map layer was built, which consisted of all the presented layers. It was published using the WMTS service in the most widely used tiling scheme, the EPSG: 3857 coordinate system with a tile size of 512x512 px. The service provided vector and raster tiles with identical content. The formats of the tiles, which we compared were:

- Vector tiles of MVT format distributed in binary format *pbf – Google Protocol buffer*, which is a neutral, extensible mechanism for the serializing of structured data (Google Developers, 2019).
- raster tiles in display format *png8*, which contains indexed colours with 8-bit samples with transparent support (Roelofs, 2006).

2.3   *Measurement*

Measurements were made on a desktop computer (RAM 16GB, Intel® Core ™ i7-7500 CPU @ 2.70GHz 2.90GHz) with a download/upload speed of 90Mb/s / 84Mb/s, ping to map server <1ms, in the Google Chrome browser. We recorded the number of rendered tiles and the duration of each rendering run in a given view. We also determined the time from the beginning of the first to the end of the last run of the single-view rendering in the map field. To measure tile rendering, we used the events from the openlayers library – *precompose* a *postcompose* and a timesheet on the browser's console. Supplementary information was given as the values of decrease of the frame rate of change of view and the calculation time of the types of operation during the measurements presented in the browser developer tools.

We then recorded the time of transfer of the tile from the server to the client (minimum, maximum and total time). To measure the time that a client retrieved a tile from a server, we used the library event *tileloadstart, tileloadend*, and t*ileloaderror*. The tiles were pre-generated and stored on the server side, and so the map preparation time for the tiles did not enter the value. There were 5 measurements for both vector tiles and raster tiles. Measurement consisted of 12 automatic view changes that simulate user activity with the map. The resulting values obtained in a particular view change step were averaged. The initial approximate value for the application launch is 8.5 in the applied tile scheme. Gradually, the map zoomed in to a zoom level 16 with a preserved area of 8.5 – 10 – 11 – 12 – 13 – 14 – 15 - 16 in the region. After the last map shift, the map zoomed out to the approximate value of 12. The single-view tiles were not queried from the server all at once due to the limiting parallel queries to the server via HTTP/1.1. This phenomenon was reflected in the fact that there was a visible time difference in obtaining the tiles, and so several rendering cycles took place until all the tiles were rendered (Figure 1).

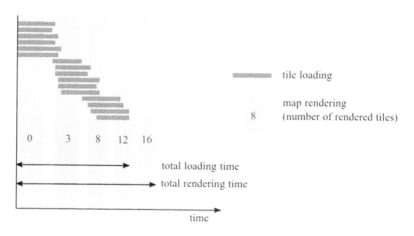

Figure 1.   Schematic of tile rendering in one view.

186

3 RESULTS

The results of the experiment were confirmed on the basis that the rendering of raster tiles takes much less time, because in the rendering, the applied raster image was created without any need of data processing. The data processing, preparation of cartographic symbols, as well as the more demanding generation of the rendering elements are a significant negative element in the applied vector tiles. Figure 2 illustrates the average duration of the course of rendering for each step of the change of view.

The differences are quite significant. Most noticeably, they are at lower zoom values where a larger area is displayed. The main factor is the volume of data in the vector tiles - the number of elements and the complexity of their geometry that needs to be rendered. A key element in the preparation of vector tiles is therefore, generalization; the evaluation of the generalization applied is described below. Other factors are the graphical representation of the data and the associated operations, such as positioning labels and detecting conflicts of overlapping. The cartographic symbolism used in the experiment is simple, since it involves generating descriptions for reference points of municipalities and along rivers with conflict detection from more demanding operations. Despite this, the necessary operations for preparation of rendering is much more difficult for vector tiles than for raster tiles. This is illustrated in Figure 3, where the average time spent in the running of the script (predominantly in the preparation of rendering) for vector tiles is more than 10 times longer. With raster tiles, painting takes up more time, which is caused by raster image decoding and image preparation. Overall, however, the calculated time is considerably smaller than for the raster tiles. It also causes a momentary decrease in frame rates for a change in view to 1-3 fps for vector tiles, while for raster tiles, it was approximately 30 fps.

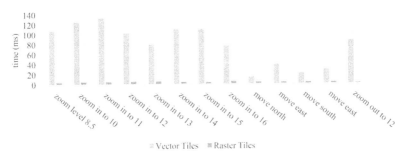

Figure 2.   Comparison of geometric rendering times for individual view changes.

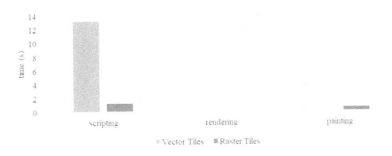

Figure 3.   Average calculated time (and its types) for the entire measured run.

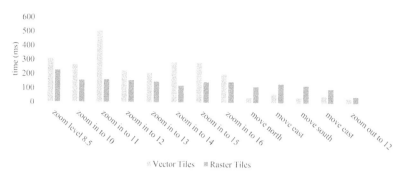

Figure 4.   A comparison of loaded times of tiles.

Table 1.   The number of vertices of the area of interest in different zoom levels.

ZoomLevel*	original	8	7	6	5	4	3	2	1
Vertices	3873	3215	2615	2069	1162	726	438	234	63

The transfer time of the tiles to the client is also directly related to the volume of data. Figure 4 presents the average tile load times for a single view. It is visibly significant that longer times at lower approach levels can be seen with more pronounced elevations at levels 11 and 14 due to the increase in volume of data in the tiles by obtaining a large number of roads of a lesser category, that is to say, buildings. In contrast, for shifts within Level 16, an area with a lower amount of elements was displayed where vector tiles were effective.

The results are influenced by the fact that vector tiles sent from the GeoServer map server do not have a degree of generalization at a level that would correspond to the map resolution where the tile is applied. This is documented in Table 1, where the generalization of the shape of the area of interest is illustrated.

To get a better idea, the displayed polygon size of the area of interest at zoom level 1 occupies an area of 5x4 px, indicating that a fraction (5-10) of the vertices would be sufficient to represent the shape in such detail. This confirms the very low extent of generalization in the creation of tiles on the map server. This fact reduced the extent of positive effect when applying vector tiles.

CONCLUSIONS

Based on the results achieved, it is possible to say that the display of data in the form of vector tiles provides significantly higher performance requirements for the client. This results in a lower image rendering speed and may reduce the comfort of work with maps. The disadvantages appear to be more pronounced in the case of insufficient optimization in the creation of vector tiles. Selection of generalization is inevitable to ensure that tiles do not contain unnecessary elements for a given zoom level. This is necessary of course, to address in detail - for extensive data layers, it is necessary to prepare rules for the gradual display of objects at different zoom levels. This will prevent a jump in the amount of data zooming in on the area. We have, however, shown that it is also very important to simplify the geometry. Here, you can save a lot of data for both transmission on the net and rendering by the client. The Mapbox specification, which describes the architecture of the vector tiles being focused on, deliberately doesn't mention the most effective way of generalizing the geometry, since only its risks are mentioned.

When displaying on smaller scales, it is preferable to use vector tiles only in the case when the data is sufficiently generalized. This problem is less significant when displayed on large scales. The great advantage of vector tiles, which needs to be perceived in the context of the experiment, is that for higher levels of zoom, the client does not need to retrieve any additional data and can render the map content from the data obtained (while keeping the rendering quality). In contrast, with regards to raster tiles, more data is required - up to 4 times more tiles at each additional level, which often cannot be pre-generated due to the disk capacity of the server.

The results presented document a specific case that depends on the architecture chosen, the hardware performance and the condition of the implementation of the web browser technology. Nevertheless, they provide an idea of the changes caused by the use of vector tiles. With the anticipated increase in hardware performance and technology optimization, especially client map libraries, the disadvantages will be overshadowed, and vector tiles will be used even more.

REFERENCES

Antoniou, V. Morley, J. Haklay, M. (2009). Tiled vectors: A method for vector transmission over the Web. In: Carswell J.D., Fotheringham A.S., McArdle G. (eds) Web and Wireless Geographical Information Systems. W2GIS 2009. *Lecture Notes in Computer Science*, vol 5886. Springer, Berlin, Heidelberg.

Ellren, P., & Gustafsson, A. (2010). *Maps from Sparse Geospatial Data Tiles*. U.S. Patent Application No 12/487,466, 2010.

Eriksson, O., & Rydkvist, E. (2015). An in-depth analysis of dynamically rendered vector-based maps with WebGL using Mapbox GL JS. *Linköpings universitet*.

Environmental Systems Research Institute, Inc. (2015). *Top Takeaways from the 2015 Esri DevSummit*. https://www.esri.com/~/media/Files/Pdfs/news/arcuser/0615/top-takeaways-from-the-2015-esri-devsummit.pdf

Google Developers (2019). *Protocol Buffers*. https://developers.google.com/protocol-buffers/

Google Inc. (2010). *Under the hood of Google Maps 5.0 for Android*. [online]. Available: http://googlemobile.blogspot.com/2010/12/under-hood-of-google-maps-50-for.html [accessed 08 May 2019].

Howell, J. R., Elson, J., & Fisher, D. (2011). *Tiled packaging of vector image data*. U.S. Patent No 7,925,100, 2011.

Mapbox (2017). *Vector Tile Specification*. [online]. Available: https://github.com/mapbox/vector-tile-spec/tree/master/2.1 [accessed 29 April 2018].

Shang, X. (2015). A Study on Efficient Vector Mapping With Vector Tiles Based on Cloud Server Architecture. *PhD diss., University of Calgary*.

Shi, N. X., & Wu, X. (2010). *Method of client side map rendering with tiled vector data*. U.S. Patent No 7,734,412, 2010. Grove, A.T. 1980. Geomorphic evolution of the Sahara and the Nile. In M.A. J. Williams & H. Faure (eds), *The Sahara and the Nile*: 21–35. Rotterdam: Balkema.

Advances and Trends in Geodesy, Cartography and Geoinformatics II –
Molčíková, Hurčíková, & Blišťan (eds)
© 2020 Taylor & Francis Group, London, ISBN 978-0-367-34651-5

The assessment of the chosen LiDAR data sources in Slovakia for the archaeological spatial analysis

T. Lieskovský & J.Faixová Chalachanová
Faculty of Civil Engineering, Slovak University of Technology in Bratislava, Bratislava, Slovak Republic

ABSTRACT: The paper deals with assessment of the quality and usability of new product of the airborne laser scanning, which is produced by The Geodesy, Cartography and Cadastre Authority of the Slovak Republic (GCCA SR). The raw data from the airborne laser scanning will be provided in the form of point cloud with density at least 5 points per square meter. After the raw data classification, the digital elevation model will be provided as the grid of elevation points interpolated from the classified point cloud data. In the domain of archaeological sites detection and spatial analysis the Light Detection And Ranging (LiDAR) method, especially from the airborne laser scanning, offers very useful and fast method for data capturing. But in the region of Slovakia the sources for such data are provided with lower density of points, or they are primarily intended for the forest management. It means, that they are captured during the period of the greatest vegetation activity, which is inappropriate for the usage in the archaeology domain. That is why the new product of the airborne laser scanning could be the interesting source of the elevation data in this domain. In this paper, the several sources of LiDAR data are evaluated within the area of interest, e.g. data obtained by the National Forest Centre as well as the data obtained within the new project of laser scanning realized by GCCA SR. The reference etalon was obtained terrestrially by surveying using total station and interpolated into the form of raster digital elevation model using ordinary kriging method. The digital elevation models were compared and tested in terms of methodology in accordance with International Organization for Standardization (ISO) standard ISO 19157:2013 Geographic information – Data quality. The paper is resulting into the recommendations for the new airborne laser scanning product usage and good practices in the domain of archaeology.

1 INTRODUCTION

LiDAR technology is a significant way to provide the data as an analytical tool in the domain of archaeology (Davis et al., 2019) and (Canuto et al., 2018). This method is used for detection of the obscure human settlements that have not yet been discovered by classical archaeological methods (Lieskovský et al., 2018). There are several products of LiDAR technology with different parameters like point cloud density, ground filtering, object's classifying etc (Anderson et al., 2005).

The aim of this study is to evaluate the main sources of LiDAR data and to assess their usability in the domain of archaeology. Two data sources were evaluated in this study: point cloud performed by the National Forest Center (NFC) and new product performed by GCCA SR. The NFC data source was obtained using LiDAR scanning in 2014. An overview of the sites scanned in individual stages is shown in (Figure 1 - left). The GCCA SR data source was obtained within the process of creation of the new airborne laser scanning product in SR and an overview of the scanned sites is shown in (Figure 1 - right).

Figure 1.   An overview of the SR territory covered by NFC LiDAR data (left) (www.forestportal.sk) and by GCCA SR LiDAR data (right) (Gálová, 2019).

2   MATERIALS AND METHODS

2.1   *Materials*

First data source which has been studied was point cloud obtained from LiDAR scanning performed by the NFC. The digital elevation model (DEM) was created from this data with resolution of 20 cm. NFC data is primarily used for the forest management and protection, so their methodology and quality are primarily oriented on these purposes. The point density of the last return in the area of interest (the Belá - Dulice settlement) (Figure 2 - left) was about 4 points per m^2 and scanning was performed during the growing season. The point cloud was filtered and 44.5% of the entry points were classified as the ground points. The density of the filtered input data remained almost unchanged in the meadow areas, on the other hand, most of the input points was filtered out in overgrown areas. It resulted into the large holes without data in these parts, which was subsequently reflected in the created DEM (Figure 2 - middle).

Second assessed data source was point cloud obtained from new LiDAR scanning performed by the GCCA SR outside the growing season. The point density of the last return in the area of interest (the Neštich settlement) was about 15 points per m^2. Input data were

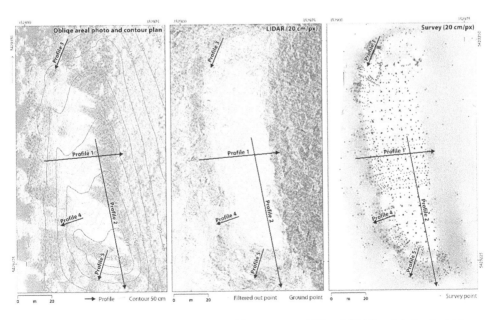

Figure 2.   NFC data source (left - orthophoto with contours, middle – LiDAR point cloud, right - geodetic survey).

191

filtered and classified with declared height uncertainty m_h = 0.15 m and positioning uncertainty m_{XY} = 0.30 m (Leitmannova & Kalivoda, 2018). It was subsequently interpolated into the DEM.

Both point clouds (NFC and GCCA SR) were transformed into or provided in the ETRS89/UTM zone 34N (EPSG: 3046) with the ellipsoid heights using the ETRS 89 datum (EPSG: 4937).

2.2 Methods

The methodology of DEM testing is based on the standard ISO 19157: 2013 Geographic information - Data quality. The data obtained by geodetic surveying using the total station were chosen as the reference etalon. The etalon was compared with both data sources: NFC LiDAR data (for the Belá - Dulice settlement and the Neštich settlement) and GCCA SR LiDAR data (only for the Neštich settlement). The reference etalon in the territory of the Belá - Dulice settlement was used in the form of the profiles (Profile 1 – Profile 5) and in the form of the collection of survey control points also (Figure 2 - right). The reference etalon in the territory of the Neštich settlement was used in the form of the collection of measured control points evenly distributed in the area of interest respecting the course of the terrain (Figure 3). The point-to-surface method (ISO 19157: 2013) was used to analyze the quality of the LiDAR data. The mean difference between the height measured terrestrially and the height obtained by the LiDAR measurement was calculated, as well as the maximum and minimum height differences, as well as the standard deviation. From the measurements were excluded the outlying values with respect to the limit value of 3σ which is commonly used in practice for this purpose. The surface-to-surface method was also used in this territory and the map of the DEM differences between the reference etalon and the evaluated LiDAR DEM (NFC and GCCA SR) was created.

3 RESULTS

Five profiles (Figure 2) were selected for the elevation validation. Profiles 1 and 2 capture the overall section of the settlement in the central part taking account the east-west direction, respectively the section of the southern part of the settlement taking account the north-south direction. Both profiles were laded through the areas of high concentration of LiDAR and

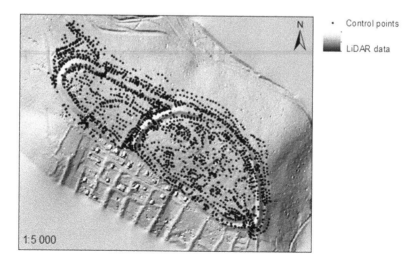

Figure 3. The distribution of the measured control points of the reference etalon in the Neštich settlement.

geodetic data. Profile 3 was chosen in the northern part of the settlement in the area with significant changes of the terrain morphology (recent interventions in the form of trenches from the World War II). It is also the area with a loss of many points after filtering the LiDAR data. Profile 4 was chosen in the area with average occurrence and profile 5 with low occurrence of the filtered LiDAR data and geodetic survey points.

The results (Table 1) indicate that the mean value of the elevation differences in the NFC LiDAR DEM in the Belá - Dulice settlement is 0.156 m. The minimum and the maximum difference values indicate the presence of rough errors (poorly filtered ground points can be assumed). But the standard deviation with value of 0.118 m points out that the presence of these rough errors is rare.

The validation of the LiDAR data in the territory of the Neštich settlement is presented on the histograms of the elevation differences between the reference etalon and the evaluated DEM – NFC LiDAR DEM (Figure 4 - left) and GCCA SR LiDAR DEM (Figure 4 - right). The rough errors were excluded from statistical calculation taking account the 3σ criterion where an observation is considered as an outlier if its least squares residual exceeds three times its standard deviation (Lehmann, 2013). The results (Table 2) indicate that the mean value of the elevation differences in the NFC LiDAR DEM in the Neštich settlement is -0.006 m and the standard deviation has the value of 0.266 m. The mean value of the elevation differences in the GCCA SR LiDAR DEM in the Neštich settlement is 0.010 m and the standard deviation has the value of 0.217 m.

Accordingly, in the territory of Neštich settlement the map of the DEM differences between the reference etalon and the evaluated LiDAR DEM was created using the surface-to-surface method. The map of the surface to surface differences for the NFC LiDAR DEM is presented in Figure 5 and for the GCCA SR LiDAR DEM is presented in Figure 6.

Table 1.   The results of the validation of the NFC LiDAR DEM in the Belá - Dulice settlement.

	Length [m]	Mean [m]	St. dev [m]	Min [m]	Max [m]	Number of points
Profile 1	46	0.153	0.115	-0.040	0.712	93
Profile 2	85.5	0.105	0.087	-0.093	0.468	172
Profile 3	17.5	0.941	0.606	0.115	1.851	36
Profile 4	13	0.176	0.055	0.077	0.276	27
Profile 5	17	0.278	0.253	-0.102	0.775	35
Survey points	-	**0.156**	**0.118**	**-0.56**	**2.34**	**1596**

Table 2.   The results of the validation of the LiDAR DEM in the Neštich settlement.

	Mean [m]	St. dev [m]	Min [m]	Max [m]	Number of points
NFC LiDAR DEM	-0.006	0.266	-0.499	0.499	2317
GCCA SR LiDAR DEM	0.010	0.217	-0.450	0.449	2564

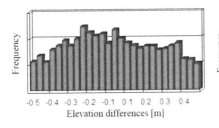 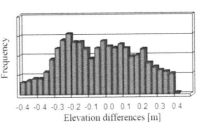

Figure 4.   The histograms of the elevation differences: NFC LiDAR DEM (left) and GCCA SR LiDAR DEM (right).

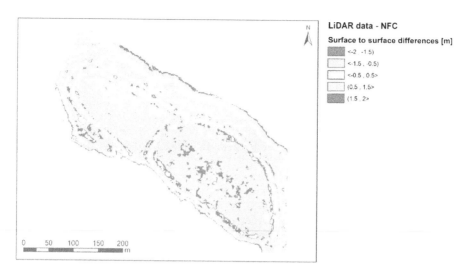

Figure 5.   The map of the DEM differences between the reference etalon and the NFC LiDAR DEM.

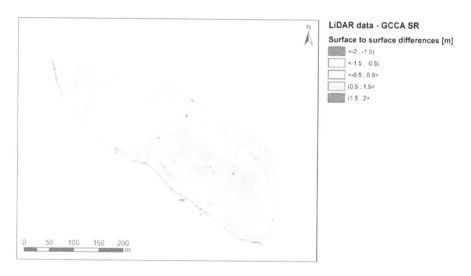

Figure 6.   The map of the DEM differences between the reference etalon and the GCCA SR LiDAR DEM.

4   DISCUSSION AND CONCLUSION

This study was intended to compare the quality of two LiDAR data sources created in the territory of SR – NFC data source and new LiDAR product of GCCA SR. The assessment was
realized in the context of archaeological analysis in according to its requirements and needs.

The results show that profiles with adequate amount of filtered LiDAR data (profile 1,2
and 4 in the Belá - Dulice settlement) can accurately capture the course of the terrain morphology. Only the marginal parts located on the border of the forest (profile 1) are partially distorted. The LiDAR data localization "above" the reference etalon can be observed in the
Belá - Dulice settlement. Also, some noise can be seen in the meadow areas. This phenomenon
can be related to dense grassland, which may not have been properly evaluated during the
LiDAR data filtering. In general morphology, systemic noise in the form of height shift may
be considered as not necessarily affecting the characteristics of distinctive terrain features. But

the question remains this effect when micromorphology is important (shallow agricultural and residential terraces, roads, etc.). Partial failure of the LiDAR data filtering in profile 5 and complete failure in profile 3 has manifested itself in the standard deviation also (Table 1). This phenomenon can also be observed in the other coniferous areas.

Within the area covered also by new LiDAR product of GCCA SR (the Neštich settlement), local maxima and minima (mounds and ditches) have the higher differences. The methodology of data quality evaluation in the case of reinforces surfaces is well elaborated and functional. But identification of the identical points is more difficult for unpaved surfaces, such as the area of tested settlements. There is also a problem of exact classification of the points into the ground category. So, in this case, the surface-to-surface method of the evaluation seems to be more appropriate.

Therefore, in the future, when designing the LiDAR scanning methodology in dense and coniferous areas, the density of the measured points should be increased, or more efficient scanners may be used, or scanning outside the growing season should be implemented (it is questionable for coniferous stands). Overall, LiDAR data can serve not only for optical or semi-automatic detection of archaeological structures such as settlements and their elements. In the case of favorable conditions, they can also be used to calculate selected morphometric parameters of archaeological structures. The quality of the LiDAR data may also be affected by the primary purpose of its collection (e.g. forestry purpose for NFC) and its adapted methodology. Therefore, in the future we propose to include the possibility of detection and documentation of cultural heritage as one of the objectives in the preparation and implementation of LiDAR scanning in Slovakia.

ACKNOWLEDGEMENTS

This work was supported by the Grant No. 1/0300/19 of the Grant Agency of Slovak Republic VEGA.

REFERENCES

Anderson, E.S., Thompson, J.A., Crouse, D.A., Austin, R.E. 2005. Horizontal resolution and data density effects on remotely sensed LIDAR - based DEM. Geoderma 2005, 132, 406–415.

Canuto, M.A., Estrada-Belli, F., Garrison, T.G., Houston, S.D., Acuña, M.J., Kováč, M., Marken, D., Nondédéo, P., Auld-Thomas, L., Castanet, C., Chatelain, D., Chiriboga, C.R., Drápela, T., Lieskovský, T., Tokovinine, A., Velasquez, A., Fernández-Díaz, J.C., Shrestha, R. 2018. Ancient lowland Maya complexity as revealed by airborne laser scanning of northern Guatemala. Science, Volume 361, Issue 6409, 28 September 2018, ISSN: 0036807.

Davis, D.S., Sanger, M.C., Lipo, C.P. 2019. Automated mound detection using lidar and object-based image analysis in Beaufort County, South Carolina. Southeastern Archaeology, Volume 38, Issue 1, 2 January 2019, pp. 23–37, ISSN: 0734578X.

Gálová, L. 2019. Stav projektu leteckého laserového skenovania SR. The Geodesy, Cartography and Cadastre Authority of the Slovak Republic. In: Computer Applications in Archaeology 18/2019, Kočovce.

Lehmann, R. 2013. 3sigma-Rule for Outlier Detection from the Viewpoint of Geodetic Adjustment. In: Journal of Surveying Engineering 139(4):157–165, 2013, DOI: 10.1061/(ASCE)SU.1943-5428.0000112. ISSN (print): 0733-9453 | ISSN (online): 1943-5428.

Leitmannová, K., Kalivoda, M. 2018. Projekt leteckého laserového skenovania Slovenskej republiky. Geodetický a kartografický obzor 5/2018, Roč. 64 (106), číslo 5, ČÚZK a ÚGKK SR, pp. 101–104, ISSN 1805-7446.

Lieskovský, T., Faixová Chalachanová, J., Lessová, L., Horňák, M. 2018. Analysis of LiDAR data with low density in the context of its applicability for the cultural heritage documentation. In Advances and Trends in Geodesy, Cartography and Geoinformatics. London: CRC Press, 2018, pp. 191–196. ISBN 978-1-138-58489-1.

STN EN ISO 19157: 2013 Geographic information - Data quality. Slovak Office of Standards, Metrology and Testing, Bratislava 2013. http://www.forestportal.sk/sites/ulzi/gis/public/Stranky/web.aspx

Advances and Trends in Geodesy, Cartography and Geoinformatics II –
Molčíková, Hurčíková, & Blišťan (eds)
© 2020 Taylor & Francis Group, London, ISBN 978-0-367-34651-5

Cadastral map quality visualization in Stupava municipality

M. Smida, Ľ. Hudecová & M. Bajtala
Faculty of Civil Engineering, Slovak University of Technology, Bratislava, Slovakia

ABSTRACT: Maps are a key component of cadastral documentation – they provide geometric and positional specification of the property and are the basis for reconstruction – setting-out of a lot boundary. They increase the trustworthiness of cadastral data. All cadastral maps are available to the public via web portals. Most of them are not based on numerical basis; map quality is not known. The fact that cadastral maps are in vector form does not guarantee their accuracy and quality. The user must be extremely cautious when a digital image of the map is interpreted. This paper features application for map quality visualization. Used methods and technics can help cadastral office to identify the places with the lowest quality of cadastral map. These localities could be suggested for renewal of cadastral documentation by new mapping priority.

1 INTRODUCTION

The cadastral maps increase the determination of all legal acts fixed to the real estate. They are available to the public in digital form using web portals. However, the most of the cadastral maps are not based on numerical origin of point coordinates, so its positional accuracy is unknown (Seidlová & Chromčák 2017), (Gašincová & Gašinec 2009). The users cannot identify its origin and quality; therefore they cannot recognize the confidence ratio of the map. This is a serious problem for all real estate owners and land market (Hanus et al. 2018). The paper features application of map quality visualization, providing considerable help to cadastral map users with their decisions on real estate market. The information about the quality of real estate can motivate owner to enable surveying activity on his property. The higher confidence ratio of registered real estate means the higher price in the land market.

The paper´s main goal is to visualize the parameters of quality in Stupava municipality that determine precision of mapping, map accuracy-homogeneity, map resolution according to scale, trustworthiness of land parcel area based on quality of detailed point (positional accuracy) and consistency of land parcel area in cadastral map with the area in descriptive data file (ownership document), (Kyseľ et al. 2018). All those source data can be obtained from files, that are available on the cadastral web sites and contain all required information. Moreover, with the help of scripts in Python programming language, all necessary data can be computed and visualized with GIS software (Smida 2016). Since there are no longer any obstacles that could intercept the plans, the thematic maps containing parameters of quality in cadastral area of Stupava can be ultimately created.

All in all, with all those methods and techniques, the critical low-quality data of cadastral maps can be simply localized and can serve as the foundation for cadastral parcel accuracy enhancement (Act No. 162/1995).

2 ANALYSIS AND DEFINITION OF PARAMETERS

Quality code of detailed points is the most important parameter for monitoring the quality of cadastral maps. However, the most of the cadastral maps are not based on numerical origin of point coordinates, so its positional accuracy is unknown. Therefore, a new system for monitoring the quality of cadastral maps was used for map quality visualization. According to the new system, the monitored parameters include: mapping accuracy, homogeneity of map accuracy, map resolution according to scale, credibility of plot area based on quality of detail point (positional accuracy) and consistency of plot area with cadastral map and descriptive data file (owner document), (Kyseľ et al. 2018).

2.1 Parameter K1 - precision of mapping

The first parameter of quality K1 is primary based on mapping technologies and its precision, that were available in the era of map creation (Hudecová 2011). Those methods are varied from old methods of mapping, for example a method of surveying table, to modern methods like GNSS and photogrammetry, Table 1.

Table 1. Precision of mapping.

Type of vector map	Code	Value
ZMVM03 with m_{xy}=0.08 m	1	1
A, THMs, THMsL, ZMVM, ZMVMv with m_{xy}=0.14 m	2	0.7
A, THMs, THMsL, ZMVM, ZMVMv with m_{xy}=0.26 m	3	0.4
SZb, SZs, SZf, THM42, THM, THML	4	0.1
VMUO	5	0

2.2 Parameter K2 - map accuracy homogeneity

The second parameter of quality K2 is applied to systematic shift of coordinates in cadastral map. Those shifts can be simple eliminated by transformations to the cadastral map, with preservation of precise coordinates in layer POINTS, Table 2.

Table 2. Map accuracy homogeneity.

Type of vector map	Code	Value
numeric VCM homogenous (without layer POINTS)	1	1
numeric VCM heterogenous (with layer POINTS)	2	0.75
non-numeric VCM homogenous (essentially VCMi)	3	0.5
non-numeric VCM heterogonous (essentially VCMt)	4	0.25
VMUO	5	0.1

2.3 Parameter K3 - map resolution according to scale

The third parameter of quality K3 is defined by level of detail of map creation and its determination of scale in map tiles. From the oldest original cadastral maps using Hungarian fathom scale, to the newest cadastral maps with SI metric system scale, the precision of the map is increased with higher resolution of map, Table 3.

Table 3. Map resolution according to scale.

Map scale	Code	Value
1000 after year 2009	1	1
1000	2	0.75
720, 1440	3	0.5
2000	4	0.25
2880, 5000	5	0.1

2.4 Parameter K4 - trustworthiness of land parcel area based on quality of detail point (positional accuracy)

The fourth parameter of quality K4 is developed by a rank of positional accuracy of the coordinate point. The positional accuracy of Tier 1 and Tier 2 points is 0.08 m, Tier 3 with edge of 0.14 m, and Tier 4 with limit of 0.26 m. Tier 5 points are not on numerical basis, so their positional accuracy is unknown. In other words, the trustworthiness of land parcel is assigned by a rank of quality of the lowest tier of coordinate point, Table 4. (Decree No. 461/2009).

Table 4. Trustworthiness of land parcel area based on quality of detail point (positional accuracy).

Positional accuracy rank	Code	Value
T1, T2	1	1
T1, T2, T3	2	0.7
T1, T2, T3, T4	3	0.4
T1, T2, T3, T4. T5	4	0.1

2.5 Parameter K5 - consistency of land parcel area in cadastral map with the area in descriptive data file (ownership document)

The fifth parameter of quality K5 is connected to the acceptable amount of difference between the area of parcel in the map and in the legal documents. The quality of the land parcel area is better, if a lesser amount of difference is registered, Table 5.

Table 5. Consistency of land parcel area with cadastral map and descriptive data file (ownership document).

Amount of difference	Code	Value
Less than $1m^2$	1	1
Less than 1%	2	0.8
1-2%	3	0.6
2-3%	4	0.4
3-4%	5	0.2
More than 4%	6	0.1

3 METHODS AND CALCULATION OF PARAMETERS

3.1 Manual data collection

The most difficult and important part of this project is the realization of data collection. For example, parameters K1, K2, K3 in Stupava municipality can be simply obtained from web portal (ZBGIS® Map Client application – the Real Estate Cadastre), (MiniGIS) or from vector cadastral map file; moreover those data remain same for every parcel in the same cadastral district. Also, it is possible to acquire parameters K4 and K5 same way, but only for smaller locations, not entire cadastral district. In order to do so, it is necessary to use more advanced calculation methods.

3.2 Automatic data collection

In order to obtain parameters K4 and K5 for 14 888 land parcels in Stupava municipality, it was a necessary to use programming scripts for data calculation. After conversion of vector cadastral map file (*.vgi) into shape file (*.shp), it was also possible to use the Python module add-ons and its console in Python programming language with attribute calculator in SQL using software QGIS. Those program tools were sufficiently helpful for all calculation.

The values of parameter K4 were calculated from vector cadastral map file, where the script had a task to localize all KLADPAR parcel map layers and its coordinate points tiers. The output data consisted of parameters were exported in *.csv file, which could be then joined with *.shp in QGIS.

The values of parameter K5 were calculated from two files. The first file was the *.vgi file, which contained all land parcel land areas in vector cadastral map. Those data are representing accurate geometric parcel areas. The second file was downloaded from web portal (CICA portal) and contained land parcel areas, which are listed in ownerships documents. The land areas of same parcels from those two files were matched with Python script and output data were also exported in *.csv file (Smida 2019).

All the gathered data are representing the quality of cadastral documentation on the date 5.3.2019. Since all parameters in all land parcels in Stupava municipality were calculated, the final approach to visualize parameters of quality was achieved.

4 RESULTS

4.1 The graph parcel quality representation

The graph is one of the methods, that can easily show in the form of a grid the relation between the parameters of quality. The best type of graph for this kind of representation is a radar graph, which can be created in various software; e.g. Microsoft Excel. The radar graph consists of 5 edges and 5 vertices, creating a pentagon. Those edges are representing 5 parameters of quality: precision of mapping, map accuracy homogeneity, map resolution according to scale, trustworthiness of land parcel area based on quality of detailed point (positional accuracy) and consistency of land parcel area in cadastral map with the area in descriptive data file (ownership document). The shape of the graph depends on values of those parameters, Figure 1.

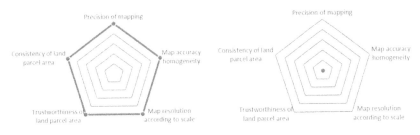

Figure 1. The examples of the best (left) and the worst (right) possible quality of a single parcel.

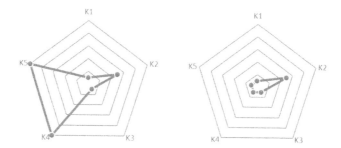

Figure 2. The examples of parcel with the best quality (left) and the worst quality (right) in Stupava municipality.

In Stupava municipality, the parameters K1 (0.1), K2 (0,5) and K3 (0.1) have constant values, therefore the graphs only differ in value of K4 and K5. The bigger the area of graph is occupied, the higher quality of parcel is registered. For the maximum possible quality of single parcel, the parameters K4 and K5 are set to value (1) and for the minimum possible quality of single parcel the parameters K4 and K5 are set to value (0.1), Figure 2.

Those graphs are ideal for visualization of parcel quality in entire municipal area and they are tracking the statistic trends after a time period of changes and activity that have influence on quality of cadastral map, for example land consolidation, cadastral new mapping etc. The graphs can be concerned as ideal addition to another forms of visualization of cadastral map quality, such as the thematic maps of quality.

4.2 The thematic map parcel quality representation

The thematic maps are considered as one of the most comprehensible and clear concepts of visualizing geospatial content (Voženílek & Kaňok 2011). They enable simple view of quality of land parcel area. The thematic map – visualization of quality in Stupava municipality was created in QGIS software with requisition of attribute database directories consisting of quality parameters in shape file. The visualization required the calculation of average value of parameters on parcel, which were set into 5 intervals of average quality. Finally, the map symbology was created to visualize 5 classes with a range from the highest quality of land parcel to the lowest quality of land parcel, Figure 3.

In the end, the created thematic map was exported with map labels, grid, legend and complementary information into vector and raster formats, in order to provide further map distribution for other users. Used concepts and content provide help for cadastral office to identify the places with the lowest quality of cadastral map in Stupava municipality. The application of the thematic map and its implementation to the public market could motivate the land parcel property owners to place an order to surveyor to perform map revision, Figure 4.

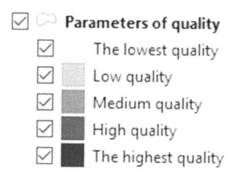

Figure 3. The map legend – the thematic map – visualization of quality in Stupava municipality.

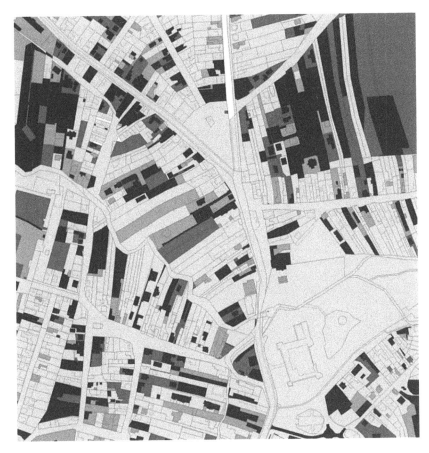

Figure 4.   The thematic map – visualization of quality in Stupava municipality.

5   SUMMARY AND CONCLUSIONS

The paper features practical implementation of cadastral map quality visualization. It defines quality parameters, which describe technical and legal aspects and origin of cadastral maps, (Table 1-5). It's showing an optimal way of calculating the parameters of quality with scripts in Python programming language. It describes various operations and progressions for qualitative visualization of cadastral documentation (Figure 1-4). The created thematic map is user friendly and easy to understand. It expresses all qualitative characteristics of a map. The thematic map can provide considerable help to cadastral map users with regard to decision-making on real estate market and with motivation of making an order to surveyor to perform map revision. All the procedures listed in this paper could be applied for qualitative visualization of cadastral documentation in every cadastral district in Slovak Republic in some conditions.

REFERENCES

Act No. 162/1995 on the real estate cadaster and the entries of ownership and other rights to the real estates (the cadastral act).
Decree No. 461/2009 of the Office of Geodesy, Cartography and Cadastre of Slovak Republic, implementing The Act No. 162/1995 Cadaster Act.
Guideline no. 12 for establishes a method for labelling vector cadastral map files. 2013. Office of Geodesy, Cartography and Cadastre of Slovak Republic, Bratislava.

Gašincová, S. & Gašinec, J. 2009. Legislative changes in the department of geodesy, cartography and cadastre of real estates since 1st September 2009. *Acta Montanistica Slovaca*, vol. 16, issue 4: 328–336.

Hanus, P., Pęska-Siwik, A. & Szewczyk, R. 2018. Spatial analysis of the accuracy of the cadastral parcel boundaries. *Computers and Electronics in Agriculture*, vol. 144: 9–15.

Hudecová, Ľ. 2011. Vector Cadastral Maps. *Geodetický a kartografický obzor* 57 (99): 243–248.

Kyseľ, P., Hudecová, Ľ. & Bajtala, M. 2018. Spatial analyse of cadastral maps. In *18th International Multidisciplinary Scientific GeoConference SGEM 2018, Conference proceedings book* 2.3, vol. 18: 567–574. Bulgaria: Albena.

Seidlová, A. & Chromčák J. 2017. Types of cadastral maps in Slovak republic and accuracy of the land area. In *MATEC Web of Conferences R-S-P, Theoretical Foundation of Civil Engineering*, Conference proceedings book, vol. 117: p 6. Poland:Warsaw.

Smida, M. 2016. The creation of the thematic map using ArcGIS Online. *Bc. Thesis*, Slovak University of technology, Faculty of Civil Engineering, Slovakia: Bratislava.

Smida, M. 2019. Selected computer techniques and tools to visualize the content and quality of cadastral documentation. *Diploma Thesis*, Slovak University of technology, Faculty of Civil Engineering, Slovakia: Bratislava.

Voženílek, V. & Kaňok, J. 2011. Methods of Thematic Cartography - Visualization of Spatial Phenomena (in Czech). Palacký University Olomouc, pp.216. Czechia: Olomouc.

CICA portal. Office of Geodesy, Cartography and Cadastre of the Slovak Republic. https://cica.vugk.sk/

ZBGIS® - Basic data base for the geographic information system Map Client application - the Real Estate Cadastre theme. Office of Geodesy, Cartography and Cadastre of Slovak Republic, https://zbgis.skgeodesy.sk/

Mini GIS portal. Office of Geodesy, Cartography and Cadastre of Slovak Republic, http://minigis.skgeodesy.sk/miniGIS/

Advances and Trends in Geodesy, Cartography and Geoinformatics II –
Molčíková, Hurčíková, & Blišťan (eds)
© 2020 Taylor & Francis Group, London, ISBN 978-0-367-34651-5

Using strain analysis to test positional precision of cartometric scanning

M. Talich, O. Böhm & F. Antoš
Topography and Cartography, Research Institute of Geodesy, Zdiby, Czech Republic

ABSTRACT: In the process of digitizing and publishing old maps, it should be kept in mind that maps created with geodetic or astronomical measurements have their own positioning accuracy. This accuracy is principally affected by an accuracy of measurement, applied cartographic projection, a map scale, used drawing method and so on. The map accuracy is important when we want to extract information about objects in maps and about their relationship. In most cases we want to know how precise the outputs are because it influences our next decisions.

Digitization of maps and map atlases should be performed on special scanning devices. A large flatbed scanner is recommended for digitizing old maps because image distortion is minimal due to the scanner's design principles. In the paper, ScannTech cartometric scanner of Proserv company with optical resolution of 800 dpi and also described experience with this large format scanner (A0+) is introduced. The necessary part of the article is devoted to testing its accuracy, which is characterized by the average position error (0.10 mm declared by the manufacturer). Mentioned are also reasons why old maps should not be digitized as documents or books are and why it is important to choose different approach. Factors that have impact on a quality and the accuracy of digitized map are also listed.

Major part of the article is given to methods for testing of an accuracy of scanners. It is proposed to use strain analysis to test the positional accuracy of cartometric scanning. The benefits of this new approach are discussed. Test sheets for doing tests are also presented. Results from long-term monitoring are also presented. These results describe behavior of scanners and show how a distortion of digital images (maps) is changing in a time.

1 INTRODUCTION

Digitized old maps are usually displayed on the internet as images, e.g. as Zoomify (zoomify.com). Another option is georeferencing the maps and provide them as Web Map Service (WMS) or Tile Map Service (TMS). Georeferenced form is usually more sought after for its practical advantages. Such maps can be used in GIS applications and are much easier to compare with other maps. To make such visual comparison feasible the positional accuracy of drawing localisation must be better than 0.4mm in exceptional cases 0.5 mm in map's scale.

Localization accuracy in tenths of millimetres in map scale can be achieved when georeferencing maps created by geodetic methods with the use of a cartographic projection. One example is georeferencing of Third Military Survey of Austrian Monarchy (1869-85) using elastic transformation with collocation. This georeferencing achieved standard error of position of 0.36mm in map scale (1:25 000) for the area of the Czech Republic (Talich et al. 2013).

To achieve such precision in georeference all cartographic properties of the map has to be taken into account. That means compensate for shrinkage of paper, take into account cartographic projection, transform coordinates and eliminate local deformations (errors of position) caused by imperfections of instruments, mapping procedures and geodetic bases. Quantity and configuration of control points have a large impact on the final accuracy. But the most basic condition is high fidelity of input data, i.e. scans of old maps.

Positional accuracy of drawing in old maps, especially large scale maps, is usually 0.1 mm. To avoid degrading map's cartographic properties it should be scanned with comparable accuracy, i.e. 0.1 mm or maximum deviation of 0.2 mm. Typically only specialized cartometric scanners satisfy this requirement. Even these scanners have to be regularly tested to ensure they still conform to the requirements.

First mentions of evaluation of scanners for their use in digitization of cartographic works come from (Carstensen & Campbell 1991). (Ho & Chang 1997) paper examines the accuracy of desktop scanners. Experimental results showed acceptable mean square error of 0.12 mm. The effort to use digitized old maps in practice led to creation of formal requirements and regulations for positional accuracy of scanners and methods of testing it. An example of such state regulations is (ČÚZK 2004).

Further research and practice showed positional accuracy is essential for the exploitation of digitized old maps. (Achilleos 2010) presented a method of estimating accuracy of digitized contours from analysis of their geometry. (Cintra & Nero 2015) devised a new way to assess accuracy from control points. At the same time research into improving the precision of cadastral maps had been underway. These efforts focused mostly on the choice of suitable transformation method usually recommending elastic transformation TPS (Felus, 2007). Other experiments tried to improve the positional accuracy by using linear elements (e.g. communication networks) for transformation (Siriba, Dalyot & Sester 2012). (Tuno, Mulahusic & Kogoj 2017) tried to improve the selection of control points to eliminate errors of position and so achieve homogeneous positional accuracy of cadastral maps.

From the above it overview it is clear that the importance of precise digitization of old maps will continue to grow in the future.

2 TESTING GEOMETRIC ACCURACY OF SCANNERS

Common desktop scanners are unsuitable for scanning maps because of the maps' usually large size. Such large maps can only be scanned on common desktop scanners part by part and the parts then have to be joined into a single image in some software. That is not only inconvenient but also introduces errors of position in individual blocks. Large format scanners, either pullout or flatbed, are a better solution. Pullout (cylinder) scanners run the map through them and scanner head remains static. On the other hand in flatbed scanner the map is static and the sensing instrument (camera) moves. Flatbed scanners are less invasive to the map and usually have better geometric accuracy.

Scanner's cameras scan the map from short distance (less than 20 cm) in rows which minimalizes imperfections of the optical system. Large format scanners typically employ several cameras with overlapping fields of vision to cover the whole width of scanned work. This yields several images (one for each camera) during the scanning process. These images are merged into a single one by the scanner's service software (firmware). This merging process, stitching, is a critical part of the procedure as it can be a source of image distortions.

Evaluation of scanner's geometric accuracy can be approached in two ways. First is assessing absolute accuracy. This is a measure of how accurately is the map reproduced, that is how big are errors of position of test points in resulting digitized map compared to paper original. The second approach is relative accuracy - measure of consistency of digitization over time and homogeneity of positional errors over the whole digitized area. In other words it is a measure of difference between repeated scans and also a measure of difference between errors in various parts of the scan.

2.1 Testing absolute accuracy

In principle absolute accuracy is evaluated by comparing digitized image of a known template with its real size and geometry. This way of testing is used for example in state regulation (ČÚZK 2004). The template is usually a regular grid of cross-shaped markers. Line widths of the markers should not be larger than 0.1 mm which corresponds to 3 pixel width in resulting image when scanning with 600 dpi resolution. The position of markers in the template has to be measured with precision in the order of hundredths of millimeters. A suitable instrument is for example laser interferometer. The template has to be made from a material resistant to thermal expansion it has to be flexible so it can be pressed against the scanning glass of a flatbed scanner or pulled through a pullout scanner. Such a template is scanned, the markers' positions (image coordinates) are measured and compared to coordinates of the markers on template. Differences in the two sets of coordinates characterize absolute geometric accuracy of the scanner.

Drawback of this method is the necessity to precisely measure the template which is a time consuming process in practice limiting the number of markers and therefore test points. Relatively low number of test points limits this method in catching potential local deformations of small areas. For example the distance between neighboring markers in the template used in aforementioned state regulation (ČÚZK 2004) is 50mm. Another disadvantage is the necessity for estimating marker centers for manual measurement of the template which can lead to errors.

2.2 Testing relative accuracy

Relative accuracy testing is a more recent method of scanner quality evaluation. It also uses a testing template of regular grid of cross-shaped markers but there is no need to measure the markers' coordinates on the template. Instead the template is scanned repeatedly and markers' image coordinates transformed into a common coordinate system are compared between multiple scans. Detection of marker position in the image can be automated with various image processing techniques. That allows much denser marker placement while avoiding laborious manual measurements. Detailed description of marker detection by image correlations, calculation of position changes and interpretation of measurements is for example in (Antoš, Böhm & Talich 2014).

A dense grid of markers is necessary to detect areas with distinctly higher deformations than the rest of the image. Such a template can be evaluated in the same way as in the case of absolute accuracy testing. This provides more detailed view of errors of position homogeneity caused by scanner and thus can be used to determine relative accuracy of individual parts of the scanning area. But this method requires precise measurement of the template to allow comparing image coordinate with reference coordinates. That can present a problem because the template can contain even several thousands of markers depending on its size and density of its marker grid. Errors in manual estimation of marker centers play a role in this method too.

3 USING STRAIN ANALYSIS FOR RELATIVE ACCURACY TESTING

Experience with practical absolute accuracy testing has shown that the results are often dependent on placement of template in the scanner. This led to a hypothesis about the existence of local deformations in areas smaller than is the gap between markers on template. When some markers lie in such an area, their positions show larger deviations. But when the template is shifted slightly so that no markers lie in these problematic areas, the marker positions are not affected by them and all deviations are small.

These small deformation areas are important for objective evaluation of scanner's relative accuracy. The only way to detect them is to use a denser template (with smaller distance between markers). That makes absolute accuracy evaluation unfeasible due to difficulties with manual reading of all marker coordinates. Relative accuracy evaluation is better method in such a case because it can be in large part automated.

A suitable mathematical tool to assess changes in relative relation between markers is strain analysis. This theory is based on continuum mechanics but in geometric sense it can be considered theory of small deformations. It describes changes in shape and size of observed objects through interpretation of repeated measurements. In this case the observed objects are markers on the template and repeated measurements are image coordinates of markers on repeated scans. Displacement vectors are a function of coordinates:

$$\mathbf{x}_i{}^\circ - \mathbf{x}_i{}^t = \mathbf{d}_i = (u_1, u_2, u_3)_i{}^T = \mathbf{u}(\mathbf{x}) = (u_1(\mathbf{x}), u_2(\mathbf{x}), u_3(\mathbf{x}))^T, \ \mathbf{x} = (x, y, z)^T$$

Where $\mathbf{x}_i{}^\circ$ (resp. $\mathbf{x}_i{}^t$) is the vector of P_i point coordinates of fundamental (resp. actual in t-time) epoch.

The strain tensor $\mathbf{E}_i$ in the P_i point is defined as a gradient of the function in this point:

$$\mathbf{E}_i = \text{grad}(\mathbf{d}_i).$$

The most illustrative indicator of the scale of geometric deformation at point P_i is total dilatation:

$$\Delta = \frac{\partial u_1}{\partial x} + \frac{\partial u_2}{\partial y}.$$

Positive values indicate extension at P_i, negative value signify compression. This allows identifying areas with largest local deformation and also the extent of these deformations. Conveniently total dilatation is invariant with relation to translation and rotation and therefore unrelated to choice of coordinate system used to measure marker coordinates. Thus there is no need to transform marker coordinates into a common coordinate system. Apart from numerical values of dilatations for discreet points, the results can also be displayed as hypsometry for the whole template.

Theoretical solution and derivation of these formulas in question may be found in a number of publications - e.g., (Szostak-Chrzanovski 2006), (Talich 2008) and (Kostelecký, Talich, Vyskočil 1994).

4   RESULT OF RELATION ACCURACY TEST OF SCANNTECH 800I SCANNER

Practical use of strain analysis for relative accuracy evaluation is demonstrated on Scann-Tech 800i scanner. The test included a total 193 of scans of a template in varied time intervals. The template was made from shrink-proof material astralon and contained 1886 markers (46 rows x 41 columns). The distance between markers in row/column was 20 mm. Time intervals between scans were 30 minutes, 5 minutes and no interval at all (save for the time it took to save the scan). The reason for varied intervals was an effort to assess effect of time between scans on geometric accuracy of scans. Scans were taken in 800 dpi resolution and 24 bit color depth.

Marker positions were detected to a sub-pixel precision using image correlation. Pixel size in this case was 0.03175 mm. Differences in image coordinates of corresponding markers on multiple scans were used to calculate total dilatations and other deformation parameters. Total dilatations were represented in hypsometry form for each scan.

Analysis of the results showed differences in reproduction of the template. Largest deformations were in places of stitching - where individual cameras' images overlap and are stitched together by scanner's software. For ScannTech 800i there are three such areas. Figure 1 shows

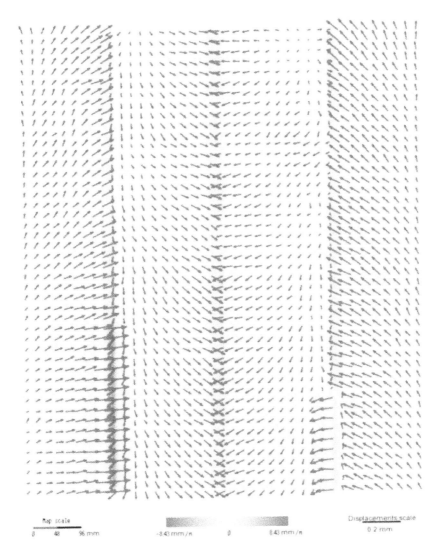

Figure 1. Image coordinate differences.

image coordinate differences between first and third scan, separated by 60 minutes time interval. The template was not moved between scans and only scanner head moved during scanning. Thus both sets of image coordinates were in the same coordinate system. Results show that the four strips taken by individual cameras are shifted relative to each other by 0.01 mm to 0.06 mm while being relatively homogeneous on their own. But the coordinate differences are up to 0.12 mm in the areas of stitching causing small areas of high local deformations. This is apparent on Figure 2 displaying total dilatations as hypsometry. Varied values of total dilatations in the same stitching areas point to imperfections in firmware joining the images from individual cameras together. Identical stitching areas show both extensions and compression on repeated scans, in some cases (in some scans) the coordinate differences were up to 0.25 mm. These small areas would likely not be detected at all using template with markers spaced 50 mm or more.

Deformations outside the stitching areas are an order of magnitude smaller. Figure 3, 4 and 5 show total dilatations for three consecutive scans with time interval of 5 minutes. There are visible distinct changes in total dilatations values between scans. Changes outside stitching areas

Figure 2.   Hypsometry of total dilatations.

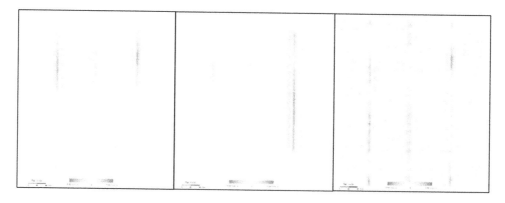

Figure 3, 4 and 5.   Hypsometry of total dilatations for three consecutive scans.

are likely caused by variations in the speed of scanner's head. This is supported by alternation of extensions and compressions in rows in Figure 5.

Results showed that time interval between scans doesn't significantly affect variations in scans of identical model. Calculations of dilatations were carried out in own special software (Talich & Havrlant 2008).

5 CONCLUSION

The paper introduces a new method of testing relative accuracy of cartometric scanners. This method uses a reference template containing a regular grid of cross-shaped markers. Template marker positions are detected with use of image correlation obviating the need for manual measurement and increasing detection precision to sub-pixel values corresponding up to hundredths of mm depending on optical resolution. Strain analysis is used to process coordinate differences between repeated scans. With strain analysis being independent on coordinate system, detected coordinates don't have to be transformed into a common system, thus eliminating errors introduced by such a transformation and improving result accuracy. In addition it can effectively process and evaluate even dense marker grids (up to thousands of markers). Thanks to that it can detect even local deformations restricted to small parts of scans. These local deformations can then be clearly represented e.g. by the hypsometry of total dilatation in the form of strain maps.

Practical tests show comparatively largest deformations occur in stitching areas (overlaps of scanner's individual cameras). Uneven speed of scanner head movement causes much smaller deformations outside the stitching areas.

REFERENCES

Achilleos G.A. 2010. Approaching a model for estimating horizontal errors of digitised contours. *Journal of spatial science*, Volume: 55, Issue: 1, Pages: 147-164, DOI: 10.1080/14498596.2010.487856

Antoš F., Böhm O. & Talich M. 2014. Accuracy testing of cartometric scanners for old maps digitizing. In: *9th International Workshop on Digital Approaches to Cartographic Heritage, Budapest, 4-5 September 2014*, 8pp. http://naki.vugtk.cz/media/doc/publikace/antos_et_all-acuracy_testing.pdf

Carstensen L.W. & Campbell J.B. 1991. Desktop scanning for cartographic digitization and spatial-analysis. *Photogrammetric engineering and remote sensing*. Volume: 57, Issue: 11, Pages: 1437-1446.

Cintra J.P. & Nero M.A. 2015. New Method for Positional Cartographic Quality Control in Digital Mapping. *Journal of surveying engineering*. Volume: 141, Issue: 3, 10pp DOI: 10.1061/(ASCE)SU.1943-5428.0000141

ČÚZK (Český úřad zeměměřický a katastrální - Czech Office for Surveying, Mapping and Cadastre) 2004. *Direction No. 32 for scanning of cadastral maps and graphic operates of former land registry. (1014/2004-22 Pokyny č. 32 pro skenování katastrálních map a grafických operátů dřívějších pozemkových evidencí)*, Praha, 20pp, https://www.cuzk.cz/Predpisy/Resortni-predpisy-a-opatreni/Pokyny-CUZK-31-42/Pokyny_32.aspx

Felus Y.A. 2007. On the positional enhancement of digital cadastral maps. *Survey review*. Volume: 39, Issue: 306, Pages: 268-281, DOI: 10.1179/175227007X197183

Ho W.H. & Chang K.T. 1997. Accuracy assessment of digitized data with randomized block model. *Journal of surveying engineering – ASCE*. Volume: 123, Issue: 3, Pages: 87-100. DOI: 10.1061/(ASCE)0733-9453(1997)123:3(87)

Kostelecký J., Talich M. & Vyskočil P. 1994. Crustal Deformation Analysis in the International Center on Recent Crustal Movements. *Journal of the Geodetic Society of Japan* - 40/4 (1994), p. 301-308. ISSN: 0038-0830. DOI 10.11366/sokuchi1954.40.301 http://doi.org/10.11366/sokuchi1954.40.301

Siriba D.N., Dalyot S. & Sester M. 2012. Geometric quality enhancement of legacy graphical cadastral datasets through thin plate splines transformation. *Survey review*. Volume: 44, Issue: 325, Pages: 91-101, DOI: 10.1179/1752270611Y.0000000011

Szostak – Chrzanowski A., Prószyñski. W. & Gambin W. 2006. Continuum Mechanics as a Support for Deformation Monitoring, Analysis, and Intrpretation, In: Kahmen, H. and Chrzanowski, A. (Eds.), *Proceedings 3rd IAG/12th FIG Symposium, Baden, May 22-24, 2006*, Baden, 9pp, ISBN: 3-9501492-3-6. https://docplayer.net/storage/95/126180372/1577541813/yo8oQFSjloYyUCkdvfd1hA/126180372.pdf

Talich M. 2008. Practical advantages of using the mechanics of continuum to analyse deformations obtained from geodetic survey. In: *Measuring the changes - joint symposia of 13th FIG International Symposium on Deformation Measurements and Analysis and 4th IAG Symposium on Geodesy for Geotechnical and Structural Engineering, LNEC, Lisbon, Portugal, May 12-15 2008*. 11pp. http://www.fig.net/resources/proceedings/2008/lisbon_2008_comm6/papers/pas07/pas07_03_talich_mc056.pdf

Talich M. & Havrlant J. 2008. Application of deformation analysis and its new possibilities. In: *Measuring the changes - joint symposia of 13th FIG International Symposium on Deformation Measurements and Analysis and 4th IAG Symposium on Geodesy for Geotechnical and Structural Engineering, LNEC, Lisbon, Portugal, May 12-15 2008*. 12pp. http://www.fig.net/resources/proceedings/2008/lisbon_2008_comm6/papers/pst02/pst02_05_talich_mc057.pdf

Talich M., Soukup L., Havrlant J., Ambrožová K., Böhm O., Antoš F. 2013. Georeferencing of the Third Military Survey of Austrian Monarchy. In: *Proceedings of the 26th International Cartographic Conference, 25. – 30. 8. 2013, Dresden, Germany*, International Cartographic Association, 19pp, ISBN 978-1-907075-06-3, http://icaci.org/files/documents/ICC_proceedings/ICC2013/_extendedAbstract/266_proceeding.pdf

Tuno N., Mulahusic A. & Kogoj D. 2017. Improving the Positional Accuracy of Digital Cadastral Maps through Optimal Geometric Transformation. *Journal of surveying engineering*. Volume: 143, Issue: 3, 12pp, Article Number: UNSP 05017002, DOI: 10.1061/(ASCE)SU.1943-5428.0000217

Advances and Trends in Geodesy, Cartography and Geoinformatics II –
Molčíková, Hurčíková, & Blišťan (eds)
© 2020 Taylor & Francis Group, London, ISBN 978-0-367-34651-5

Author Index

Advances and Trends in Geodesy, Cartography and Geoinformatics
ISSN: 2643-2420
e-ISSN: 2643-2404

1. Advances and Trends in Geodesy, Cartography and Geoinformatics I
Proceedings of the 10th International Scientific and Professional Conference on Geodesy, Cartography and Geoinformatics (GCG 2017), October 10-13, 2017, Demänovská Dolina, Low Tatras, Slovakia
Edited by Soňa Molčíková, Viera Hurčíková, Vladislava Zeliznaková & Peter Blišťan
ISBN: 978-1-138-58489-1 (HB)
ISBN: 978-0-429-32702-5 (eBook)

2. Advances and Trends in Geodesy, Cartography and Geoinformatics II
Proceedings of the 11th International Scientific and Technical Conference on Geodesy, Cartography and Geoinformatics (GCG 2019), September 10-13, 2019, Demänovská Dolina, Low Tatras, Slovakia
Edited by Soňa Molčíková, Viera Hurčíková & Peter Blišťan
ISBN: 978-0-367-34651-5 (HB)
ISBN: 978-0-429-32702-5 (eBook)